c|net

Do-It-Yourself Home Video PROJECTS

24 cool things you didn't know you could do!

Troy Dreier

New York Chicago San Francisco
Lisbon London Madrid Mexico City
Milan New Delhi San Juan
Seoul Singapore Sydney Toronto

The McGraw-Hill Companies

Cataloging-in-Publication Data is on file with the Library of Congress

McGraw-Hill books are available at special quantity discounts to use as premiums and sales promotions, or for use in corporate training programs. For more information, please write to the Director of Special Sales, Professional Publishing, McGraw-Hill, Two Penn Plaza, New York, NY 10121-2298. Or contact your local bookstore.

CNET Do-It-Yourself Home Video Projects:
24 Cool Things You Didn't Know You Could Do!

1234567890 QPD QPD 01987

ISBN: 978-0-07-148933-1
MHID: 0-07-148933-9

Sponsoring Editor
Judy Bass

Editorial Supervisor
Janet Walden

Project Manager
Madhu Bhardwaj (International Typesetting and Composition)

Acquisitions Coordinator
Alexis Richard

Copy Editor
Bill McManus

Proofreader
Bev Weiler

Indexer
Kevin Broccoli

Production Supervisor
Jean Bodeaux

Composition
International Typesetting and Composition

Illustration
International Typesetting and Composition

Art Director, Cover
Jeff Weeks

Cover Designer
Jeff Weeks

About the Author

Troy Dreier is a freelance tech writer based in the New York City area. He's a regular writer for WebVideoUniverse.com, where he gives tips and covers trends in online video, and is a frequent contributor to CNET, *Laptop Magazine, Computer Shopper,* PDAStreet.com, and Intranet Journal. He also writes a weekly consumer technology column for *The Jersey Journal* newspaper. Contact him at tdreier@gmail.com.

Contents

Foreword

Of all the personal technology revolutions we've seen in the last decade, the pinnacle of them may be desktop video editing. With an inexpensive camcorder, software, and modern PC, you can produce videos that just ten years ago might have only been possible with professional equipment costing more than the average home!

But all too often, powerful technology creates so much content and choice that we get overwhelmed and do nothing with any of it! (How many long, unedited video tapes are sitting on a shelf at your place?) One thing technology doesn't change is the human attention span, and watching flabby raw footage of *anything* is just no fun. That's where this book comes in, framing your thinking on what to do and showing you how to do it.

This book assumes you have learned how to use your camera and software—then we bring it all together. From simple techniques that create magic, like "green screening," to outlines for productions that show you how to create a message that moves and compels. And once you finish a couple of these projects, you may want to offer them up to the world through the unbridled spread of online video sharing.

With all this we bring today's technologies for personal expression full circle: digital photos, personal web sites, blogs, and digital-music collections are all one fraction of the expression now easily accomplished through home video editing. Ready, set, ACTION!

Brian Cooley
CNET Editor-at-Large

Introduction

Thanks for picking up this book, which I think you'll find is different from the other home video titles on the shelf. While the others dutifully take you through the steps of shooting video, importing your work, adding titles, and so on, this book is organized by projects. The idea is not just to show you how to use your camera and software, but to show you specific things you *can* do. Some of the projects are practical—like sharing your creations or archiving your work—while others are designed for the whole family—like creating a video holiday card or a family newscast. Still others are meant to show you how to do things you never thought possible with home video, like shooting a stop-motion movie or creating your own green screen effects. Along the way, I've scattered lots of tips and advice, so you can get the most out of your time. I hope there are several projects that sound fun to you, and that you have as much fun using this book as I did in putting it together.

A few words of advice if you're just picking out equipment: The trend in today's cameras is to offer HD video and built-in hard drives, or cameras that record video directly to a DVD with a built-in burner. While those cameras make playback and archiving simple, they make editing almost impossible. If you care about editing your work (and if you've chosen this book, you probably do), then choose a MiniDV camera instead. It's not the only format that makes importing and editing easy, but it's the best choice for most consumers, and MiniDV tapes are inexpensive.

While you're at the camera shop, a few accessories will take you far. Pick up a tripod, an extra battery, an external microphone, and a camera bag. You'll need them for the projects in this book.

There are several decent video editing programs on the market, but I chose to use three for the examples in these projects, since I couldn't cover them all. Apple iMovie HD is a great, easy program, and it's free with every Macintosh computer as part of the iLife Suite. For Windows computers, Adobe Premiere Elements is by far the best program out, although it can be challenging for people new to editing. If you'd like something simpler, try Ulead VideoStudio, which offers a good combination of features and ease of use.

Once you're set up, have fun with your camera and don't leave your footage unedited. Look to the accompanying video tutorials on CNET (at http://diyvideo.cnet.com) for even more home video advice.

Happy shooting.

Troy Dreier

Part I

Filming

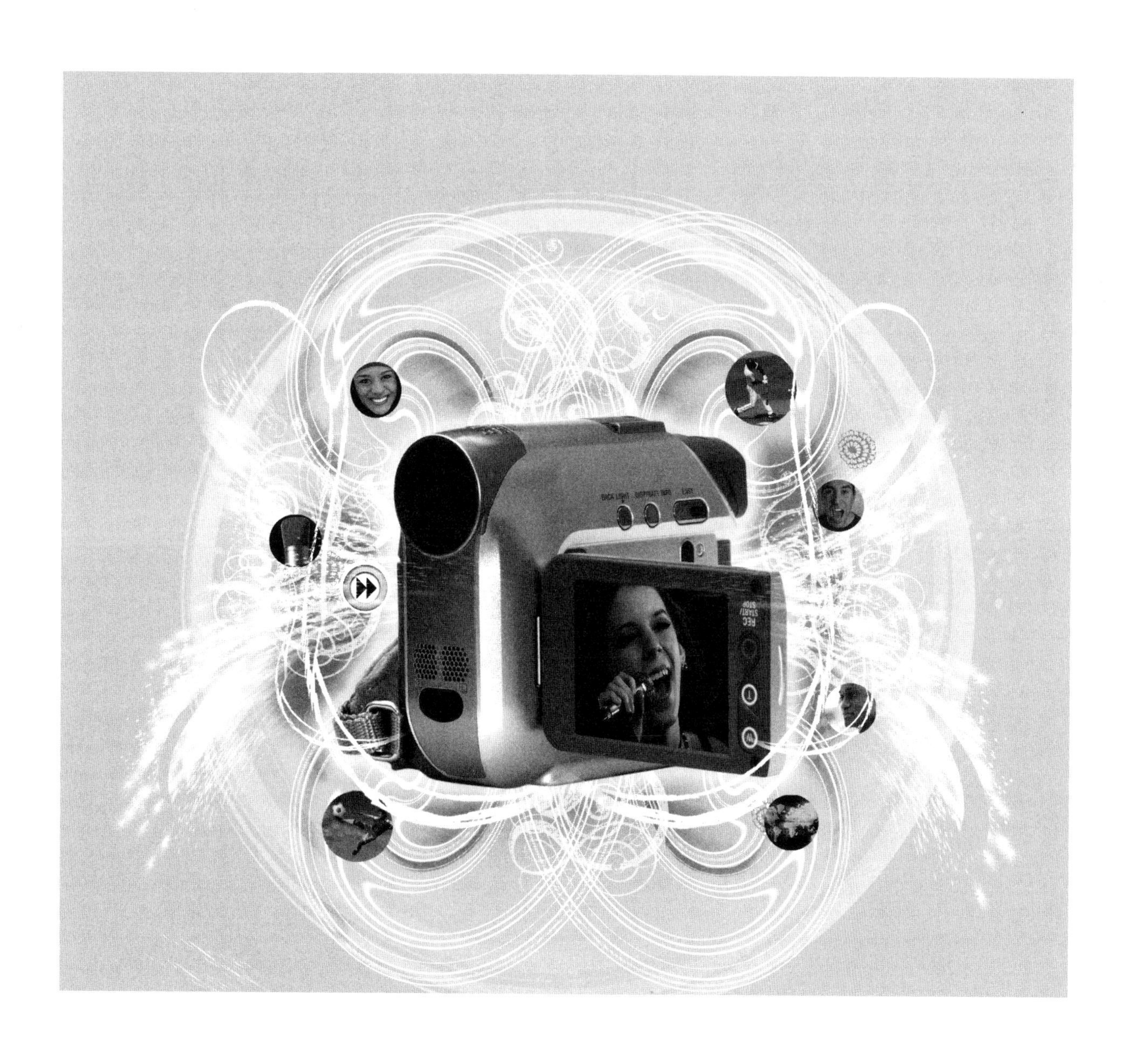

Project 1

Make a Video Birth Announcement and Record Your Kids Growing Up

What You'll Need

- **Digital video camera (any)**
- **Home computer with DVD-burning drive**
- **Blank DVDs**
- **DVD cases and mailers (available at any office supply store)**
- **Video editing software**
- **Cost: $10 to $140 (this and all other projects assume you already have a video camera and a computer)**

Our first project is actually two projects grouped together, but they're small and deal with the same idea: capturing your kids on video and sharing it—now or in the future. For a lot of video camera users, the camera only comes out when it's time to record a child's birthday, soccer match, or choir concert. If that describes you, you'll like these introductory projects, which offer simple but creative ways to record your child.

Create a Video Birth Announcement

Sending birth announcements for a new baby is a quaint custom that you don't see as often as you used to. That's too bad, because people like to get an announcement with a picture, telling them the newborn's name, weight, and birthday. Without too much work, new parents can revive the custom and give it a digital twist.

Step 1. Keep Your Video Camera Charged and Ready

Video cameras take hours to charge, and you won't have that kind of time once contractions start. If you want to capture your new son's or daughter's first moments in the world (or if you want to capture the delivery itself—although that's probably not material for a video birth announcement), be sure your camera is permanently charged.

Also, have your camera bag (see Figure 1-1) packed and ready to go as the date approaches. Besides the camera itself, you'll want to pack:

- A spare battery, also charged and ready
- Your camera's power cord (just in case)
- A tripod, so that dad can get in some of the shots, as well

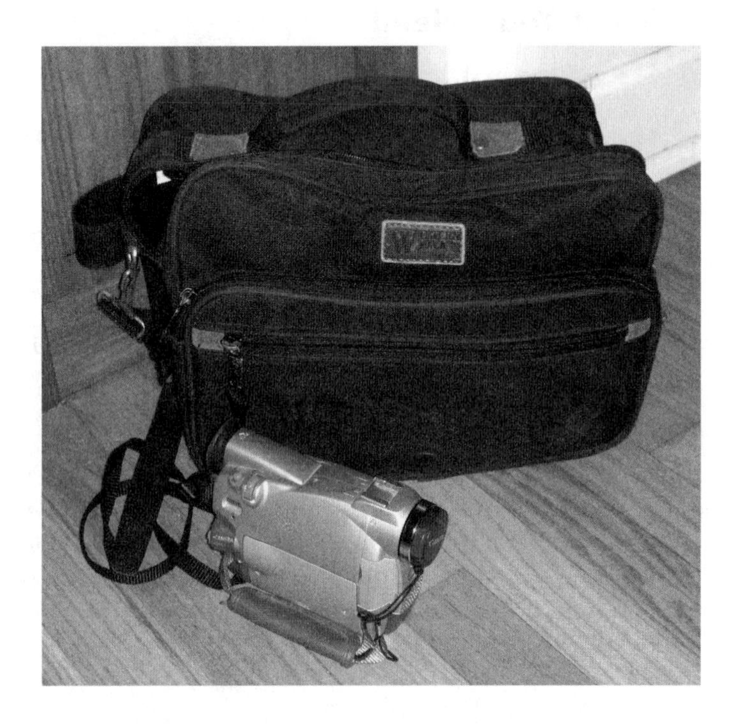

Figure 1-1

Have a camera bag standing by and ready.

Step 2. Get a Mix of Shots

When the big day finally arrives, have fun with the video recording and try to get several different kinds of shots, in several different settings. While the purpose of the video birth announcement is to show off your new bundle to the world, you can also tell a story. Creating a narrative will make the video more fun for the recipients.

tip *The harsh lighting of a hospital waiting area and the dim lighting of family birthing areas can play havoc with your results. Set your camera's white balance to correct for challenging conditions. First, put your camera in program mode, as opposed to automatic mode. This lets you change more of the settings. You'll probably find the program mode switch just inside the LCD compartment, accessible when the LCD is opened. Call up the onscreen menu and then select white balance. While aiming your camera at a perfectly white wall or a white sheet of paper, click to set the white balance. Your camera will recalibrate for the new lighting conditions. If you're shooting a white sheet of paper, zoom in first so that the paper fills the entire screen.*

Consider getting shots like these for your announcement:

- **The drive to the hospital** This is easier to do if you're taking a cab or if a third party—a friend or family member—is driving. If it's just mom and dad, presumably dad will be driving, and holding a camera will certainly be the furthest thing from mom's mind (especially if it's the middle of the night). Get a quick shot if you can; it will likely be the establishing shot for the video, setting things in motion, and will make the video more fun for your viewers. Capturing the panic and excitement you're feeling might seem crazy at that moment, but you'll be glad you have it years down the line.
- **Checking in at the hospital** You don't need to spend a lot of time showing yourself filling out forms. This is a transition shot, showing one step of the journey.
- **Waiting** For most couples, the next step is to wait in a hospital birthing room until mom's contractions are close enough together. It could take hours, especially if this is the mother's first child. This is a good time to set the camera on a tripod and talk about your feelings. How will having this child change your life? What do you wish for him or her (or them)? Introduce any close friends or family members who have come to the hospital (see Figure 1-2).

tip *Want to get in on the action, instead of holding the camera all the time? Nearly every video camera lets you flip the LCD screen so that people in front of the lens can see what's being shot. The image flips automatically when you flip the screen. This lets you tape yourself while you monitor the results at the same time.*

- **The delivery** If the hospital uses a separate delivery room—rare nowadays—the person holding the camera will have no trouble taking a video camera in to capture the actual birth. If you choose to record the delivery, you'll want to save most of the footage for your own private home movie, not the video birth announcement (it's a bit too...exciting, let's say,

Figure 1-2
Waiting for the baby? Fill time by getting a few shots of family members who have come by.

for a homemade birth announcement). Be sure to get a shot of the doctor holding up your new child as it takes its first breath of air. If dad is too squeamish to be a part of the delivery, shots of him nervously waiting for news are also charming.

- **The first feeding** After the birth, nurses will clean and dress the child, and then take him or her back to the parents in their private or semiprivate room for baby's first feeding. This is the gold at the heart of your video, the magical, restful time that everyone's been waiting for. Show mom and baby together, and capture every gurgle or cry. More friends and family members may have arrived by this time. Get shots of them holding your new little bundle. Be sure to address the camera and tell the name of your new family member, the birth time, and the birth weight, since that's what a proper birth announcement does.

tip *Planning on showing your video on a widescreen TV? Then shoot your video in widescreen mode. Look for a widescreen or 16:9 setting on your video camera. 16:9 is the aspect ratio for widescreen TVs.*

How long you should make your shots depends on the type of video you have in mind. Keeping things brief is a good idea. People are more apt to enjoy a homemade video if they can get through it quickly. After all, you don't want to risk boring people when you are announcing one of the happiest events in your life. Also, shooting short takes means that you have less to edit out later on, and makes it more likely that you'll finish your project and get it in the mail quickly.

On the other hand, you may want to shoot everything that happens on the big day and then create a much longer video for your own home history. This will make isolating the best parts for your video birth announcement a little more of a chore, but you'll have the satisfaction of saving every moment from that day for future viewing.

Step 2a. Get Shots of Other Additions to the Family

Video announcements of new additions to the family aren't just for couples who have expanded their families in the traditional way. Creating a video announcement for a child who arrived through adoption or from a surrogate mother is just as important. In fact, your friends and family might appreciate it more, since they'll be curious about how the process worked.

Consider getting shots of these various steps:

- Interviewing prospective birth mothers or adoption agencies
- Traveling to a foreign country to adopt the child
- Waiting inside or outside a delivery room
- And, of course, holding your new child for the first time

Sometimes a new addition has four paws and fur. If you're going to add a puppy or kitten to the family, take your video camera along when you go to pick out your new pet. Your friends and family will find it just as adorable to get a video birth announcement for a new pet (see Figure 1-3). Here are some shots to consider, if getting a pet:

- Calling breeders or shelters to check availability
- Interviewing yourself about what kind of pet you'd like to get, and what you plan to name it
- Visiting the breeder or shelter to view the available animals
- Playing with the various animals before you choose one (or before one chooses you)
- Taking your new dog or cat home for the first time

Step 3. Edit Your Footage

As a new parent (or pet owner), you'll have lots more important things to do than edit your footage, which is a good reason to keep your segments short while filming. You want to get the video birth announcement out while the news is still current, but if the job becomes too much of a chore, that video is likely to remain on your MiniDV tape or your computer's hard drive until your child is old enough to vote.

Figure 1-3

New puppies deserve video birth announcements too.

note *If you don't like the idea of discarding any of your footage, try this trick: import all your footage from your camera to your computer's hard drive, and then save the video file to a CD or DVD before you begin editing.*

Editing is a mental hurdle for many video hobbyists, for a variety of reasons:

- You don't know which software to choose or all the packages seem too expensive.
- You've bought a professional or semiprofessional editing program, and using it is too difficult.
- The idea of making cuts to your video and shaping the piece is so foreign to you that you are afraid that you'll do the wrong thing.

caution *If you plan on editing footage later on, don't save it as a DVD, which puts it in a finished format. Leave the file as is and burn it as a data file, as you would when backing up files. Then you'll always have an archive of the untouched material, if you want to work on it days or years down the line.*

There are several good video editing programs on the market. If you need help choosing, look to this book's introduction, which helps you get started with equipment. I'll use three programs primarily when showing and discussing edits in this book:

Apple iMovie HD, Adobe Premiere Elements, and Ulead VideoStudio. iMovie is part of the iLife suite of creative applications that comes free with all new Macintosh computers. You can also buy the suite separately. It's an excellent program that lets beginners create attractive results. Adobe Premiere Elements and Ulead VideoStudio both run on Microsoft Windows PCs only. Premiere Elements is a light version of Adobe's professional video editing program. It offers more features than any other consumer-level editing program, but also takes longer to learn. If you feel up to the challenge and you don't mind reading a manual, it's the one to get. If, on the other hand, you like things uncomplicated and you wouldn't read a manual for money, Ulead VideoStudio is the way to go. It's an elegantly simple program that makes movie editing easy.

caution *If you're not an experienced video editor and you're struggling with a professional-level program that you got from a friend—such as Apple Final Cup Pro or Adobe Premiere Pro—you'll probably find video editing a tedious chore and you won't do it. Delete that program at once and get yourself a manageable consumer-level program.*

When you're done shooting, it's time to edit your work. Follow these steps, if you've never edited video before:

1. Plug your camera into your computer to transfer the footage. Your camera will have come with the necessary USB or FireWire cable.
2. Open your video editing program. If your camera is connected and switched on, you'll see an option to import video. If you have a hard disk–based camera, you need to use the special software that came with the camera for the import.
3. Making your first cuts can be a challenging decision. Start by organizing the material into scenes, if this didn't happen automatically when you imported the footage. Load your clip into the timeline at the bottom of the screen and place the playhead cursor where you want the scene cut. Select the Cut command. In iMovie, for example, you can make a cut by right-clicking and selecting Split Video Clip at Playhead from the pop-up menu, or by choosing Edit | Split Video Clip at Playhead.
4. Once you've got your different scenes, trim away dull material, such as before people start talking and after they stop talking. Send any footage you're not using to the trash can.

tip *Trim away any footage where your camera is moving, or else you're liable to give your viewers motion sickness. Keep the moments when your camera is in position and holding a steady shot.*

5. You should start to see a basic storyline take shape. Further refine your material by creating different moments in each scene. For example, if you're interviewing different people who have dropped by to see the baby, isolate a few sentences with each person, and trim away the time when you're moving from one person to the next. Don't worry if the cuts seem too harsh; you'll soften them later with transitions.

6. When you've broken the scenes down as much as you'd like, check the length of your project. Ten to fifteen minutes is a good length for a birth announcement, and it's always better to keep viewers wanting more. If you're tempted to go longer, remember that this is just an announcement of the new child; the viewers will have to visit to see the real thing. If you need to cut additional material, search for scenes that run too long and trim them down in the same way.
7. Once you've chosen your scenes, you'll want to add transitions. Your video editing program might have hundreds of fun transitions that let one scene explode or swirl or drip into the next. Avoid these. A birth announcement should have a gentle feeling, so use simple dissolves. Find the transitions in your editing program. In iMovie, for example, you choose the Editing option on the bottom right and then click the Transitions button on the top right. All programs let you preview transitions before you add them, so click on different ones to see what they do. When you find one you like, simply drag it into the timeline, between two scenes. Use only one or two transition styles in a project, to keep a unified feeling throughout.

Avoid startling scene transitions unless you're going for comedy. You want people watching your scenes, not the transitions (see Figure 1-4).

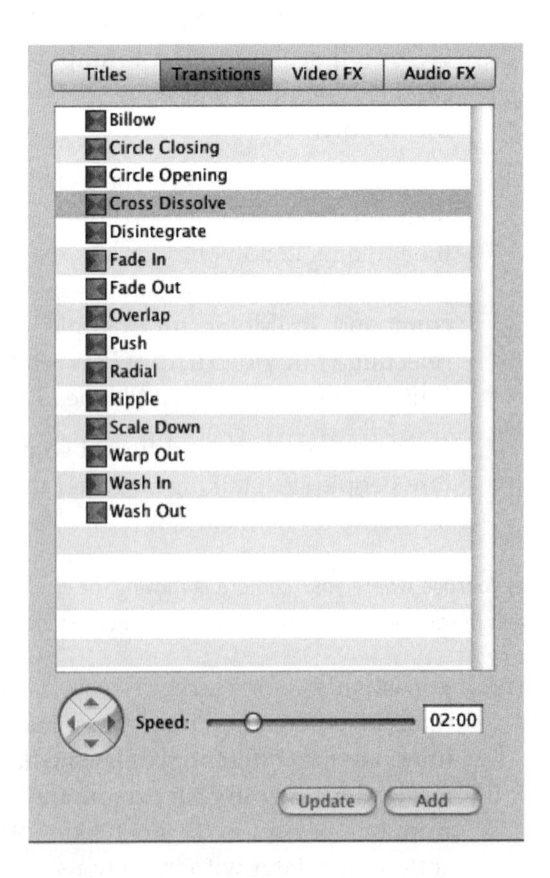

Figure 1-4

Keep transitions simple for a video birth announcement.

8. After adding transitions, you'll want to add a little music. For a commercial video project, you would need to get the rights to the music you use, but for a personal project, you're safe taking songs from your CDs or digital music collection. In iMovie, click the Media button in the lower-right corner and then select iTunes from the window at the top. You'll see all the songs in your iTunes library. Find the song you want and drag it into the audio track of the timeline below. Select songs that add to the mood you're going for, such as songs sung to babies or songs with the feeling of lullabies. Add music to quiet scenes, such as the opening and closing shots, or shots of the parents holding or feeding the new baby. Putting music under shots where people are speaking can be distracting, so if you do so, keep the volume down. If you're using iMovie, play with the "rubber band" volume level on the audio track, which lets you easily raise and lower that track's volume in key areas simply by dragging that line up or down.
9. Your final step should be to add titles and, if you wish, end credits. Your editing program will have dozens of different title styles, so choose something childlike that suits the mood. In iMovie, select the Editing button on the lower right and then click the Titles button on the top right. You'll see dozens of title styles. Click any of them for a preview. Options under the list of title styles let you input text, change the lettering color, and direct how the titles will move. You can also select a font and set the title's speed. When you have the titles just right, drag the title to the correct place on the editing timeline.

Step 4. Create the Final Package

When you've gotten the movie into the shape you want, the hard work is done, but you still need to burn it to a DVD.

Some video editing programs include a DVD creation program, but not all do. Apple iMovie doesn't include DVD creation, but the iLife suite includes iDVD, which works seamlessly with iMovie. Adobe Premiere Elements and Ulead VideoStudio both include DVD menu creation and disc burning.

If you're using iMovie, pass your video to iDVD by clicking the Share pull-down menu and selecting iDVD. iDVD automatically opens with your finished movie in place. If you're using VideoStudio, an all-in-one program, select the Share tab along the top and then select the option for creating a disc. This launches the DVD wizard. If you're using a separate DVD program, choose the Export option to save your movie to your computer's hard disk, and then launch the DVD program and import the file you just created.

Making a DVD menu is a lot of fun, because DVD programs have premade menu templates that you can select to make a professional-looking DVD. Whatever program you're using, you're sure to find a template with a baby motif (see Figure 1-5, for example).

The finished DVD can include more than just the movie. You can import all your still photographs of the newborn to create a slideshow that will play on the TV. In iDVD, for example, you do this by selecting Project | Add Slideshow (see Figure 1-6) and then importing the pictures you want.

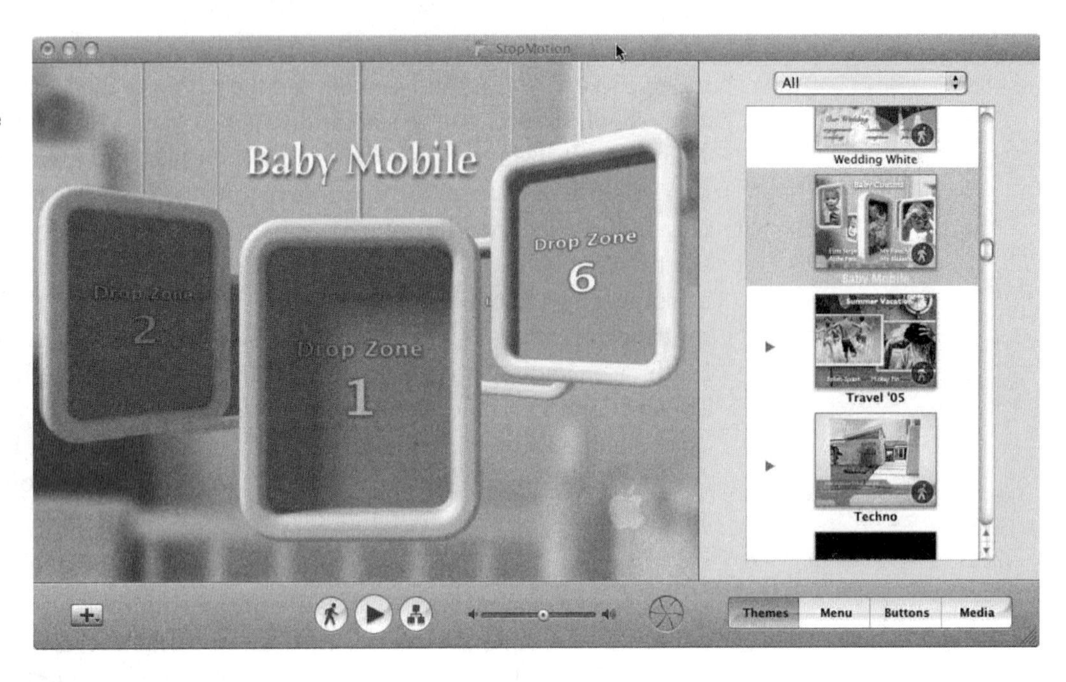

Figure 1-5

Apple iDVD has a great menu template for new-baby DVDs.

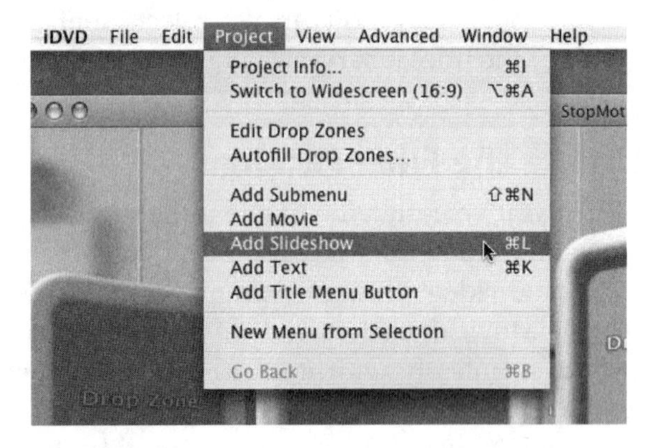

Figure 1-6

Adding a slideshow in iDVD, or most DVD-creation programs, is a snap.

tip *Another reason to keep your video short is so that burning it to DVDs takes less time. You could share your creation in other ways, such as uploading it to a web site or e-mailing it (for more about online distribution, see Project 22), but an occasion like the birth of a child requires something with a more finished look, and people will appreciate having a keepsake.*

Most video editing packages contain a simple cover-printing program, so you can upload a digital photo of your new child, or take a still photo from your video, to create the cover, then add some titles. Note that while Apple iDVD can create sensational DVD onscreen menus just right for a birth announcement, it doesn't offer a cover-creation application. Apple users should instead use the editing tools in iPhoto to crop an image to the right size and then print it out.

Keep the project manageable so that you can have your DVDs finished and out the door in a small amount of time. One evening should be about right.

Send out your DVDs in stiff DVD mailing envelopes, such as the type shown in Figure 1-7. Then, go spend some time with that beautiful new baby!

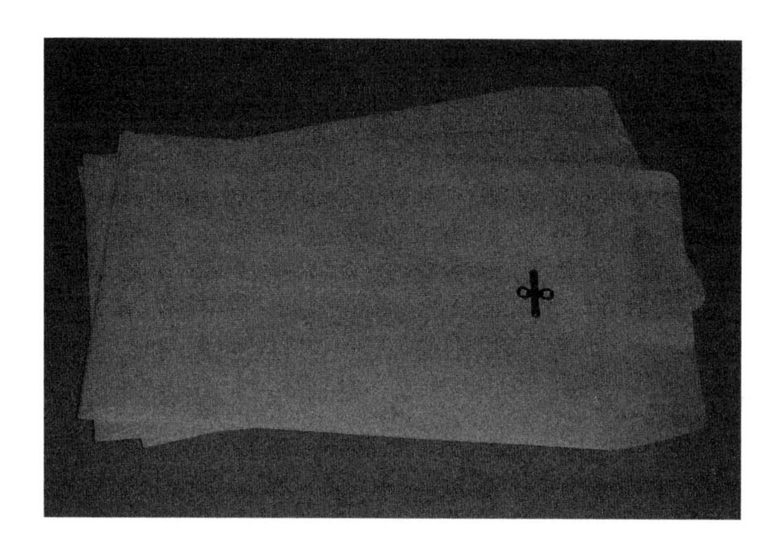

Figure 1-7
Have stiff mailing envelopes ready for your finished DVDs.

Create a Birthday Milestone Video

In the days before inexpensive video cameras, parents measured their children's growth by notching their heights on the same wall every year on the child's birthday. Now you can use a camera to create a much more personal marker.

The goal of this easy project is simply to create a video in which you interview your child once a year, on his or her birthday, answering the same questions. Think of it as raw material for a future project, since you probably won't be editing it for a while.

Step 1. Create a Questionnaire

The interview doesn't need to be long, since you'll eventually be stringing several of them together, but you need a list of a few questions to ask every year. Consider the following suggestions:

- What do you like to study in school?
- What do you want to be when you grow up?
- Who are your best friends?
- What are your favorite toys?
- What do you like to do when you're not in school?

- What's the best thing that happened to you this year?
- What was the best movie you saw this year?
- What was your favorite birthday gift?

A kindly spokesparent will have to help out with a one-year-old's interview.

Step 2. Choose a Place for the Interview

When picking a place to conduct your interview, think about choosing a spot you can use year after year. Then the viewers can focus on the answers your child is giving, and not on an ever-changing background (although it might be amusing to track how the décor in your home changes over the years). Choose a well-lit, airy space, such as an entryway staircase. Avoid shooting the interviews outside, since you can't always count on good weather.

tip *Try to conduct each year's interview in the same spot, and keep the interview free from distractions. That will make the final results easier to watch.*

Step 3. Store Your Videos

This is a long-term project, so you'll want to store your interviews in a secure way.

caution *Don't just toss the MiniDV tapes into a shoebox or keep the uploaded video files on your hard drive. Tapes can go bad, and formats like MiniDV tapes likely will not be around by the time you want to do something with your clips.*

The best option is to save your footage to your hard drive and then burn the video file to a data CD or DVD, storing the raw video file. This will make the footage easier to access when the time comes to create a larger video. Video formats shift over time, and the format you use may not be a popular option a decade or two down the road, but if you have the raw video files, they'll be easier to reformat into the newer system. For more on archiving your videos so that they'll last, see Project 24.

caution *If you do decide to save your video on MiniDV tapes, be sure you or someone in your family doesn't overwrite them. MiniDV tapes have a slider button on one end that puts them in either Record or Save mode (see Figure 1-8). Put used tapes in Save mode so that people will get an error message if they try to use them for recording.*

Step 4. Plan for the Future

Maybe it'll be a bar or bat mitzvah, a wedding, a confirmation, or graduation, but one day when your child is grown, you'll finally want to pull out the various interview videos and make something of them.

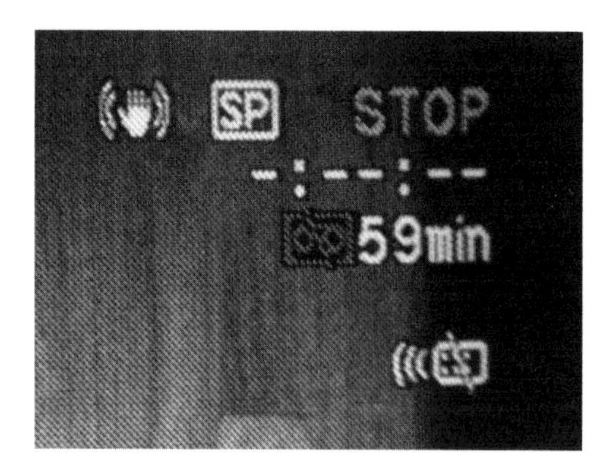

Figure 1-8

If you try to use a MiniDV tape that's set for recording protection, you'll get an error message.

note *Remember that you'll need a way to show off your video. If you plan to play it directly from a notebook computer to a large-screen television, make sure you have the necessary audio and video cables. If you need a projector to show your work, you can rent one from a local computer supply house.*

Edit your videos together into a work that's about 10 to 15 minutes long and you'll have the perfect crowd-pleaser for a special event. Even though you might have much more footage that you can use, keep the video on the short side. Crop the various clips tightly, so that one answer quickly moves into the next.

caution *Don't be tempted to try to save money by projecting your video onto a white wall. Walls are rarely perfectly white, and imperfections on the surface can ruin the look of your video. Buy or rent a screen instead.*

The video camera is a favorite toy for new parents, but you don't need to settle for boring footage that only a parent could love. Edit your video into a finished package and you'll have results that you're proud to show—and that your friends and family are happy to watch.

Project 2

Create a Family Newscast

What You'll Need

- **Digital video camera (any)**
- **Home computer**
- **Vocal and lavalier microphones**
- **Video editing software**
- **Cost: At least $140 for the microphones and software. The price might be much higher if you opt for higher-quality mics.**

Our second project is a fun activity parents can do with their kids. Creating a newscast requires shooting several scenes with family members playing several different parts. Everyone in the family gets a chance to join in. You can even create disguises and have people play more than one role.

Along the way, you'll learn about different types of microphones, how you can use them in the shoot, and why you'll get more impressive results if you do. Sure, you can get by creating a newscast with only the on-camera microphone, but your video will look—and sound—much better if you try different mics. Plus, this project offers microphone tips that you can use in other video projects.

The Different Types of Microphones

You can enhance the quality of your family newscast (and other video projects) by using different types of microphones, several of which are described in the following list. Although this project does not require that you use any of these mics, using a variety of mics will make your results more professional and will be more fun for the kids.

note *Most consumer video cameras have both microphone and headphone ports, typically hidden by a movable flap. If you can't tell how to attach a microphone to your camera, consult the manual that came with the camera.*

- **Vocal mic or unidirectional mic** This is the most common type of microphone, and it looks like any standard handheld mic that a singer would use on television (see Figure 2-1). *Unidirectional* means that it records sound from one direction. Expect to pay between $15 and $100.

Figure 2-1

A vocal mic is a fairly inexpensive investment that will make your videos sound much better.

- **Omnidirectional mic** Omnidirectional mics are used to record meetings or events, where sound is coming from several different directions. Expect to pay $40 to $100. This project doesn't use this type of mic, but it deserves mention as an alternative to the unidirectional mic.

Why pay more for a microphone? Because the better quality the mic, the better it is at recording low sounds without creating an unpleasant hiss.

- **Lavalier mic** This is the type of microphone that clips to a person's lapel. It's small and unobtrusive, and is useful in interviews. Lavaliers come in both wired and wireless varieties. Wireless versions cost more, but are far more useful because the person being recorded isn't tied to the camera. Lavaliers are generally omnidirectional. Expect to pay $25 to $200 for wired versions, or $300 to $500 for a wireless kit.
- **Bluetooth mic** As of this writing, this is a brand new category, with only one microphone in it (the Sony ECM-HW1). But the idea is so useful that copycats will certainly come around before this book is published. A Bluetooth mic gives wireless performance at a more affordable price. The ECM-HW1 has a range of 100 feet, and only works with certain Sony video cameras. It's far larger than a lavalier mic, though, so it would be hard to disguise in a shot. Expect to pay $100.

tip

If you want to attach more than one microphone to a camera, you'll need to plug them into an audio mixer first. You can find low-cost mixers for under $100.

- **Shotgun mic** This is the microphone equivalent of a zoom lens, letting you direct the sound you want recorded, and grab it from a distance. By doing this, you block out surrounding sounds that could ruin your shot. Some models attach to your camera, but professionals often have a second person hold the mic. There's a huge range of shotgun mics available, so you could pay anywhere from $50 to $700.

Starting Your Newscast

When you've got your equipment ready, it's time to begin shooting your family newscast. If your kids are going to be the newscasters, have fun with costumes before you start. Put them in matching blazers or style a wig like a local news anchor.

Whenever you're working with a microphone, be sure to plug a pair of headphones into your camera's headphone port and monitor the results while you shoot (see Figure 2-2). It's easy to forget to switch the microphone on, and if you're listening in while you shoot, you'll discover such small errors at a time when it's easy to correct them.

Figure 2-2
Headphones are a must when using external microphones.

Step 1. Set the Scene

Create a desk where your newscaster can read the news. This can be any table or desk with a sheet over the front. Have your children create a poster-sized logo to dress up the set, giving the name of their newscast.

Step 2. Bring in the Boom

Create a boom microphone. The preceding equipment guide didn't mention boom mics because you're going to make one yourself. First, get a long stick that can support the weight of a vocal microphone. A yardstick is perfect. Next, tape the mic cord so that the microphone itself dangles loose from one end, as shown in Figure 2-3. Put more tape around the middle and other end of the stick, to keep the microphone cord in place.

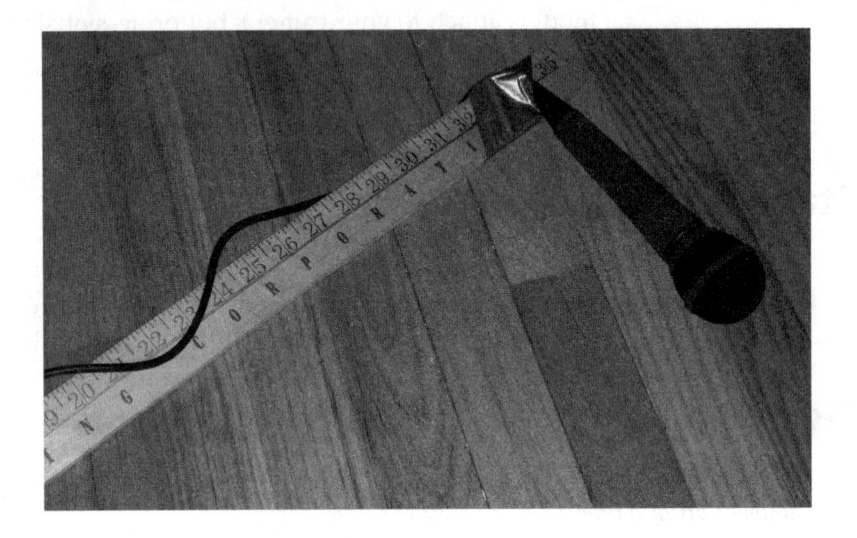

Figure 2-3
Create an impromptu boom microphone.

Step 3. Hold that Mic

Choose a tall family member to be in charge of the boom mic, since it will need to be out of the way during the shots, just above the top of the shooting area. Position the camera in a steady place or use a tripod, if you have one. Don't try to hold the camera by hand, because your muscles will get tired and shakiness will show up on the video.

Step 4. Deliver the News

Have your newscasters read from a prepared script or just improvise. It's up to you and your children how you want to work. If preparing a script sounds like too much work, feed them story ideas before each shot. They could report on family vacations, for example, or projects they're doing in school.

The Person-on-the-Street Interview

Newscasts aren't filmed only in studios. Sometimes a correspondent asks people on the street their impressions on various hot-button topics. You can bring this spontaneity into your newscast with a simple handheld mic (see Figure 2-4).

Microphones can pick up wind noise when used outside, which can be surprisingly loud in the final recording. To avoid having wind noise ruin your shot, be careful to shield the mic from any wind.

Step 1. Prepare the Reporter

Again, have fun with costuming. Put your correspondent in a trench coat and a fedora. Create a "Press" badge to put in the hat's brim.

Figure 2-4

Have your reporter interview people on the street.

Step 2. Set Up the Mic

Plug in a vocal mic. Since you probably are using a wired mic, the camera operator needs to stay close to the interviewer. It may take some practice to avoid pulling on the cord from either direction.

Step 3. Steady Your Cam

Keep the camera steady. In a person-on-the-street interview, the interviewer might move around, but the camera should stay fairly steady. If your camera races along next to the interviewer, you're going to give viewers motion sickness.

Step 4. Ask the Hard Questions

Conduct some interviews. It's up to you whether you want to stage the interviews with friends or family, or actually interview neighbors on your street. Conduct a few brief interviews and keep them all on the same topic, so it feels like a real news segment. Instruct your interviewer on good microphone techniques, which means holding the mic up close to his or her mouth when speaking, and moving the mic so that it's near the other person's mouth to catch the reply.

The One-on-One Interview

Not all interviews are done on the street. People such as government leaders and celebrities are brought into the studio for a one-on-one interview. You can do this at home by creating a set with two chairs and no desk. Put the chairs near each other, angled partly toward the camera.

tip *If you're using wireless microphones, be sure you have extra batteries on hand. Wireless mics can burn through batteries quickly, so it's better to be prepared.*

Figure 2-5
Lavalier mics clip on to a lapel or collar.

Use lavalier mics (see Figure 2-5), which clip on to the lapels or shirts of the people talking for this segment. Lavalier mics give a more informal feel to a sit-down interview, since the people talking don't need to worry about positioning a microphone.

Is your lavalier mic creating a rustling sound? That could be the fabric around it rustling against the mic. You can fix this with a little well-placed tape to cover the offending bit of fabric. This works well when the mic is hidden behind a lapel, and the tape doesn't show.

Step 1. Set Up Your Camera

Position your camera with a tripod, to hold it steady, or prop it up at the right height on a table or tray. Your subjects will be staying in place, so you don't need to worry about moving the camera.

If you hear a crackling sound from a wireless mic through your headphones, it means the mic's battery is low. Change it before you do any more filming.

Step 2. Conduct the Interview

Have your interviewer introduce the guest and begin with a few introductory questions, to set the subject. Once the interview is rolling, the interviewer can ask more pointed questions. You can go in several different creative ways with a one-on-one interview, from interviewing a parent about curfews and allowances, to having the guest play a character, perhaps a blow-hard politician. Perhaps the interviewer can help solve the mystery of who took the last cookie in the bag, or where a lost game piece is located.

The Exposé

A "gotcha" segment is standard on local news shows nowadays, where an investigative reporter protects consumers by uncovering some kind of scam. Have fun with the idea by uncovering a scam artist on your newscasts.

Step 1. Prepare the Microphone

You need the vocal mic for this segment, as the reporter will hold the mic throughout. Remember that the camera operator needs to stay close by if you're using a wired mic.

Step 2. Plan Your Shots

Shoot at least two different shots for an exposé. In the first, have the reporter show the viewers a problem in the area, or a person who has been ripped off somehow. Use your imagination to create a fun con act for the reporter to uncover. If a victim is involved, have the reporter briefly interview the victim. If the scam didn't involve a victim, have the reporter show where the scam took place and describe what happened.

Step 3. Execute the Sting

For the second scene, have your reporter confront the person being exposed. You might show the person realizing on camera that he or she has been caught. You could also show the con artist running from the camera as the reporter presents the facts. Have fun with it.

Time for Sports

What would a newscast be without a sports segment? If you have a child who is involved in playing sports, this is a great way to mix in actual footage from a match. As another example, you could shoot footage at a family outing to the bowling alley (see Figure 2-6).

Figure 2-6
Any local game makes great sports footage.

Step 1. Shoot Some Sports Footage

Record some type of actual sporting contest to use in the segment. It could be anything from a young child's t-ball game to a parent's after-work softball match. You'll be providing commentary on the action later, so think about what you want to show and say while you're shooting. You'll get more targeted results that way, instead of simply shooting everything on the field.

While recording, you can leave yourself voice notes about what you want to say in the narration, because you're going to strip the sound out later, anyway. If you think of an especially good line, say it while you're recording so that you'll remember it later on. Mention players' names, so you don't forget them later.

Step 2. Create an Introduction

Return to the anchor desk set that you used in the first part of your newscast and have the sports reporter sit next to the anchors. Let one anchor introduce the sports reporter and hand the segment over. The sports reporter should give an introduction to the game being reported on, and then segue to the video.

Step 3. Add the Sports Video in Editing

You'll add the pretaped sports video in the right place during the editing process. Most of the sports reporter's job will be done as narration, not on-camera work.

Step 4. Mic Your Computer

When you're ready to record the narration, take your standard vocal mic and plug it into the microphone or audio-in port on your computer. If you've never attached a microphone before, you'll want to adjust your computer's settings to make sure it knows where to grab the audio from. In a Windows computer, open the Windows Sound Hardware Test wizard from the audio controls, and follow the onscreen instructions, as shown in Figure 2-7. For a Macintosh computer, open the Sound system preference and select the correct sound input source from the list of sources on the screen.

Figure 2-7
The Windows Sound Hardware Test wizard.

Step 5. Prepare the Software

You might need to tell your editing software where to grab audio information from as well. In Adobe Premiere Elements, for example, you open Edit | Preferences | Audio Hardware and then select the audio input source.

Step 6. Prepare the Clip

Call up the sports clip in your editing program and place the position indicator where you want narration to begin.

Step 7. Record the Narration

Click the microphone button to call up the voice recording window. Play the clip in the main window, and then click Record on the audio controls. Have the reporter narrate the onscreen action and put a lot of energy into it, just like a real newscaster does. Don't get frustrated if recording the narration doesn't go smoothly the first couple of times. Speaking over video and highlighting the action might look easy when you see a professional do it, but it can be challenging.

Getting the right sound out of your narration takes work. Practice getting the booming voice of a seasoned TV newscaster. Always have a glass of water nearby when recording narration, as your throat can get surprisingly dry or scratchy when speaking into a mic. Better yet, add some honey or lemon to that water.

Mute your speakers before you begin recording or else you'll get horrible feedback as the mic picks up the sound from the speakers and creates a screeching sound loop.

Step 8. Block the Attached Audio

After your narration is recorded and automatically tied to the clip, delete or mute the audio track of the sports match itself. If you don't, it will be difficult for viewers to hear what's being said.

The Weather Report

A good newscast needs a weather segment, too, but that requires green-screen editing, and we're not going to tackle that here. See Project 11 for instructions on how to do it.

When you've finished shooting, you'll only need a little basic editing to trim the clips and create your first family newscast. Working with microphones isn't nearly as complicated as it sounds, and it makes your videos a lot more watchable. Once you've tried shooting with a mic, you'll never want to rely on your camera's built-in microphone again.

Project 3

Create a Video Valentine

What You'll Need

- **Digital video camera (any)**
- **Home computer**
- **Video editing and disc-burning software**
- **Blank DVDs**
- **Cost: $110**

We're going to take a brief break from the family-oriented Projects 1 and 2 and do something romantic with this project. Valentine's Day is a great time for homemade gifts—especially if you haven't paid off all the presents you purchased in December—and making a video valentine is the perfect way to show your love to your partner and make a keepsake that he or she will want to watch over and over again.

This project talks about several ways to add atmospheric lighting to a video project. It's a natural tie-in for a valentine video, but you can use these tips for other kinds of videos as well. Best of all, these lighting tips don't require you to buy any special lights. You'll improvise everything you need from materials you probably have around the house.

The Video Valentine

A video valentine is a DVD of romantic talk and special memories intended just for your beloved. We'll look at several different clips you can create to add to your DVD. Feel free to add your own ideas or modify the ones described here.

Atmosphere is important on a romantic DVD, so we'll explore different ways to create a mood. As you set up your own shots, think about not only what you're going to say or do in your video, but how you can best create a romantic atmosphere.

The Introduction

Every story has to begin somewhere, and a DVD of romantic clips might begin with a simple introduction, where you tell your beloved how you feel about him or her.

For this clip, you're going to be well lit and looking as good as you can, seated and speaking directly to the camera. To show yourself in the best possible light, you'll need to learn a little about three-point lighting.

Three-point lighting is the standard method used by portrait photographers and videographers to light subjects in a flattering way. The idea is that you want lighting to hit the subject from more than one direction, to avoid flattening out their features.

Step 1. The First Point of Light

Set up a stool or chair where you'll record this first segment. Position the camera in front of you, with a tripod if you have one. If you don't have a tripod, you can simply set the camera on a table or tray.

Your primary light source is called the *key light*. It should be in front of you to the right or left, midway between the camera in front of you and a point alongside you. In other words, the light from the key light should hit you at a 45-degree angle. This is your brightest light. If you have track lighting, such as the type shown in Figure 3-1, you can adjust one of the lights to hit where you're sitting, or you can even use sunlight coming through a window, if the weather is cooperating. Position the key light so that it lights you from above.

Figure 3-1

Use the lights you have on hand instead of buying expensive professional lighting.

Step 2. The Second Point of Light

On the opposite side of the camera from where you've placed the key light, you'll place your *fill light*. The fill light sits slightly closer to the camera. Its purpose is to eliminate shadows on your face caused by the key light. Position it a little lower than the key light. You can improvise with a desk lamp, set to light you from the right angle. Take a few test recordings as you set up, to make sure the lights are positioned in the best way.

tip *Paper lanterns provide great lighting for video, because they cast a lot of light and the light is diffused through the paper shell.*

Step 3. The Third Point of Light

Finally comes the *backlight*, which is set behind the subject and a bit to the side. Like the other two points of light, the backlight should hit you from an angle. The backlight's job is to light the back of your head, and it makes your image "pop" by creating a clear separation between you and the background. Again, a desk lamp placed on the ground and positioned at the back of your head will do the job.

tip *Thinking about investing in an on-camera light? Don't do it. And try not to use a built-in light, if your camera has one. Those lights are harsh, and over-light nearby subjects. They also come directly at the subject, flattening the subject's features. If the quality of your video matters, you're better off setting up other lights.*

Step 4. Deliver Words of Love

Now that you're all set up, turn the camera on, position yourself between the lights, and deliver your message of love. This is an introductory video, so besides telling your partner how special she or he is, you might want to introduce the other clips on the DVD.

tip *Every time you change the lighting for a scene, you should adjust your camera's white balance. Doing so keeps you from looking yellowish under certain lights and bluish or greenish under others. It gives your camera a baseline to work from, so that it can filter out unflattering lighting effects. For instructions on how to adjust the white balance, see Project 1.*

Special Places

For your second clip, you're going to go on a field trip. Take the camera along to several places in your town (or elsewhere) that have romantic or sentimental meanings for the two of you.

Step 1. Shoot Outdoor Scenes During the Golden Hour

For this project, you'll probably be shooting several outdoor places, so you're going to have to rely on natural lighting. But that doesn't mean you have no control over the light. Try to shoot any outdoor shots in the first or last hour of daylight (See Figure 3-2).

Figure 3-2

Try to shoot outdoors during a golden hour.

Photographers call these times the "golden hours." The sunrise or sunset light is tinged with pink and gold, and subjects shot during these times simply look much more impressive. Also, light comes at you from a shorter angle during these times, so you won't get harsh shadows across your face, as you might when the sun is overhead.

Visit several different places with romantic meanings, such as the place where you first met, had your first date, and so on. Record yourself saying a few words at each place. You don't need to make each segment long; just tell what each place means to you.

tip *Professional photographers use light reflectors to angle sunlight onto a subject's face and remove any shadows. If you're having a friend assist you on the shoot, pick up a reflector for about $20 at a photography store, or create an improvised one by folding tin foil over a sheet of cardboard (see Figure 3-3).*

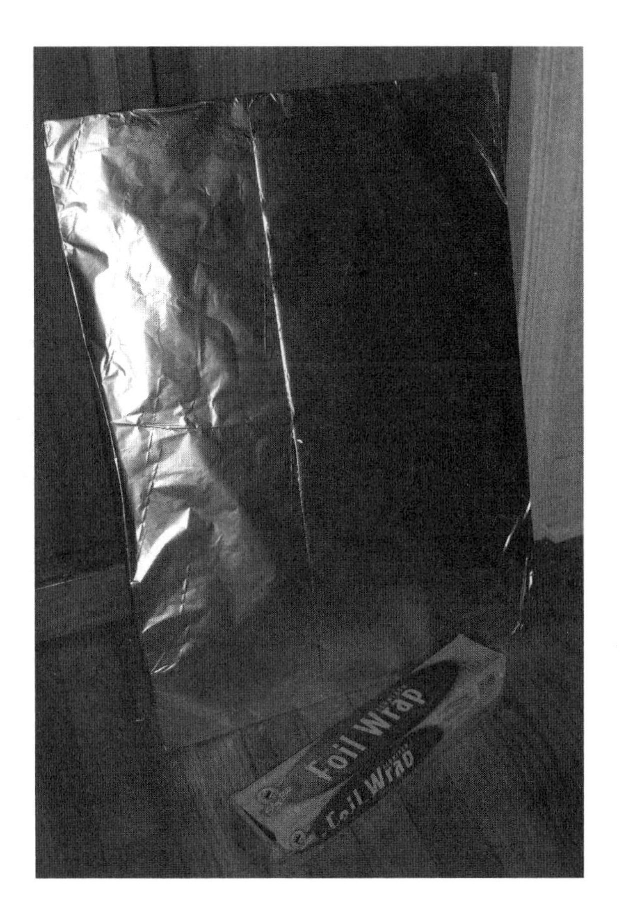

Figure 3-3

Create your own budget reflector with cardboard and tin foil.

Pillow Talk

This is a great clip to add to your video valentine if your partner frequently travels for business, staying overnight in other cities. If he or she has a notebook computer along with a DVD player, you can deliver your good-night wishes via DVD.

Step 1. Set Your Scene

You're going to film this in your bed, so you might need to wait until your partner is away on business to record it. If you film at night, you can get the bedroom looking dark, so it really looks like a night shot. If that's not possible and you need to record during the day, pull down the shades to block out as much light as possible.

Step 2. The Soft Touch

Prepare the camera by stretching a piece of plastic wrap across the lens and holding it in place with a rubber band (see Figure 3-4). Take a tiny amount of petroleum jelly and rub it across the plastic wrap. This is an instant softening filter, to blur the edges a bit and give you a more romantic look.

Figure 3-4

It's easy to create a glamorous softening filter with only plastic wrap and petroleum jelly.

Set the camera on your partner's pillow, so you can speak just as you would when the two of you are in bed. Get dressed for bed, as you normally would.

Step 3. Night Light

Experiment with a few different lighting styles, to see what looks the best on your camera. If your video camera has a night mode, turn the room lights down and switch on the night mode. See how that looks through the camera's LCD screen. If the results are natural, stick with that, but if they look greenish, try turning on only a reading light or a small flashlight. The idea is to keep light to a minimum, so that it feels like bedtime.

Step 4. Say Goodnight

Deliver a sweet and romantic goodnight that your beloved can watch just before turning in at a hotel. It doesn't have to be anything long, just a few words from home to make your traveler feel less alone.

Bathing by Candlelight

Adding loads of candles to a scene creates an instant atmosphere of warmth and romance, and you don't need to spend a ton on expensive lights to achieve it. In this valentine clip, you'll enjoy a soak in a bubble bath while surrounded by as many candles as you can get your hands on, all shining like stars. This is a good place in the DVD to deliver a seductive message. We'll stop right there; this is a G-rated book.

Step 1. Create a Romantic Glow

Find all the candles in your house and place them around the tub. Light them. Add a bubble bath mixture and fill the tub.

Step 2. Make a Star Filter

While you're waiting for the tub to fill with water, create an improvised star filter on your video camera lens with a pair of nylon stocking (see Figure 3-5a). Stretch a stocking tightly across the lens and hold it in place with a rubber band (see Figure 3-5b). Now, any light shot through the lens will take on a star effect, with lights glowing like stars (see Figure 3-5c). You can use this effect any time, but it looks especially good with lots of candles around.

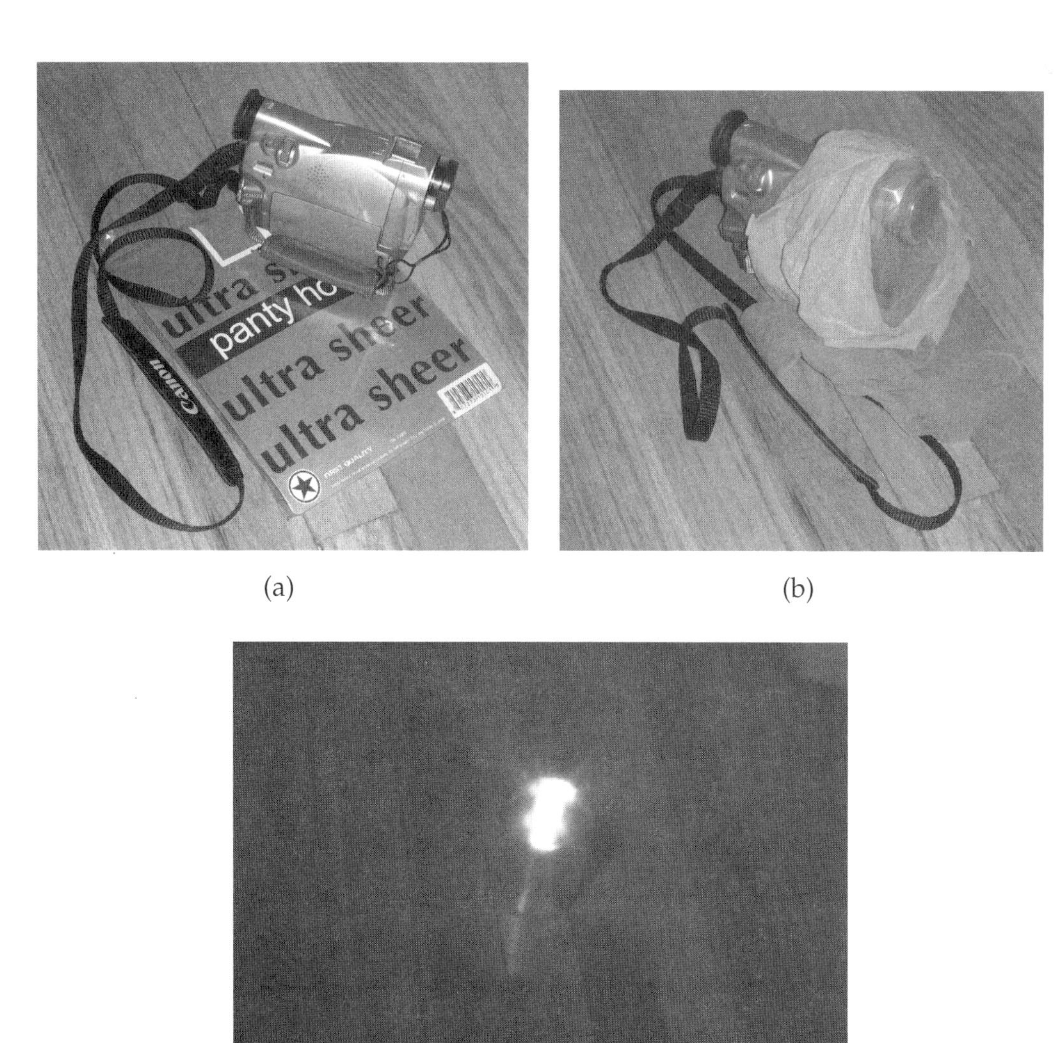

(a) (b) (c)

Figure 3-5

Nylon stockings can create an easy star filter.

Step 3. One Steamy Recording

When you're all ready, step into the tub and record your message. Don't worry that it needs to be long. Even a minute-long recording is fine.

Your Personal Top Hits

Not every clip on your video valentine needs to be a recorded video. Try making one entry on the DVD called Jukebox or Favorite Dances. When your partner selects it, a special song plays while only a picture of you two displays on the screen. Ask your partner if you can have this dance, and then lead him or her to an area where you can slow dance. Add in as many songs as you'd like to this part of the DVD. Who doesn't like a little romantic slow dancing?

The Video Massage

Finally, we've got one last surprise to add to your valentine. Create an entry on the DVD called "Healing Touch." When your partner clicks it, have it play a soothing new-age track. Explain that this is tranquil music to get a massage by. Then, tell your valentine to lie down so you can massage him or her. Make sure the music lasts for an hour or two, since if you do a good job, you just might get a massage in return.

Making the DVD

Once you've got all your clips recorded, it's time to edit them and save them to a DVD.

Step 1. Import Your Footage

Open up your video editing program and connect your video camera. The software will automatically give you the option of importing video from the connected camera.

Step 2. Break Up Your Clips

You aren't making one long movie with all of these clips; instead, you are making a variety of short clips. Cut out only the section of video that you need for each clip and save that in the program's timeline, which stretches across the bottom of the screen. Save the rest of the footage temporarily in the program's stored clips area. In Apple iMovie, for example, click the Clips tab in the lower-right corner, if it's not already selected, and you'll get a row of empty slots above it where you can store clips that aren't needed yet, as shown in Figure 3-6.

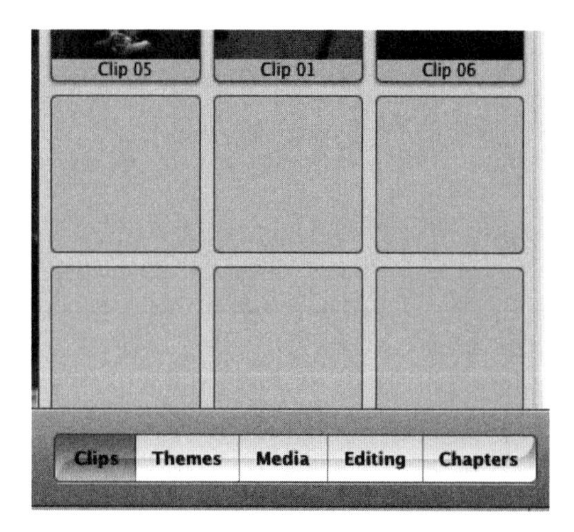

Figure 3-6

Store footage that you don't need to use yet.

Step 3. Edit to Perfection

Use the editing tools to remove any unwanted footage from each clip. Drag the play cursor in the timeline to the area that you want to remove, and then click to split the clip in two. Then, discard the unused piece.

Step 4. Save Those Mini Movies

You need to save each finished clip separately, so that you can clear the timeline and work on a different clip. Save the finished clips to your desktop or video folder, so you can easily find them when it's time to make the DVD.

Step 5. Create the DVD

When you've finished with your last clip, switch to your DVD-burning program, or just switch to the DVD-creation menu if it's built into your editing software. You first need to select a menu style for your DVD. If your software offers one with a valentine theme, it's a natural. Apple iDVD has a great romantic menu with a heart outlined in rose petals, shown in Figure 3-7.

Step 6. Add Your Clips

Import your clips one by one, and arrange the onscreen menu so that your clips are in the order you want, with the introductory clips first and so on. Add some music to your DVD menu, by adding a track to the audio section of the software. This song will automatically play when the DVD is inserted.

Figure 3-7
Select a DVD menu template

Step 7. Stuff in a Slideshow

The DVD menu gives you an opportunity to add one more element to your video valentine, a slideshow. Select digital photos from your collection that will help your partner remember all the good times you've had together, including vacations, births, celebrations, and more.

Step 8. Burn that Valentine

When you're done creating the DVD, you're ready to burn it to a blank disc (See Figure 3-8).

Congratulations—if you've followed this project and made a video valentine, you've spent hours creating a keepsake from the heart, one that means more than a store-bought card. Your partner will definitely appreciate the time you took to create something unique.

Figure 3-8

The final product

Project 4

Create a Vacation Movie Your Guests Will Really Enjoy

What You'll Need

- **Digital video camera (any)**
- **Home computer**
- **Video editing and disc-burning software**
- **Cost: $100**

There are two times when a video camera always gets pulled out of the closet and put into action. The first is when your kid is appearing in a school play, dance recital, choir concert, or similar activity. The second is when it's time for a family vacation.

There isn't a whole lot you can do to dress up the first occasion—you can only film what's in front of you—but there's lots you can do when shooting a vacation. Have a little fun with your camera and try some unusual techniques. After all, you're not shooting a documentary. It's not your job to dutifully stroll behind the family and record everything that goes on. If you have a little fun with your recording and take a few chances with unusual shots, your final film will be that much more enjoyable.

But remember, no one wants to watch unedited footage. If you want people to watch your movie and have a good time doing so, you need to edit it when you get home. Luckily, good editing isn't really that hard, and this book can teach you how.

tip *You're sure you have enough MiniDV tapes for the trip? How about picking up a spare battery before you go? A little planning up front can save you from missing shots later on.*

Shoot the Unusual

You know the typical drill for vacation shooting: you and your family arrive at each day's activity, where you pull out your video camera and record everything you all see, say, and do. At the end of the trip, you've got a little video diary of your highlights. But how about shooting some unexpected scenes, to change your footage from a highlights reel to a compelling movie? Here are some suggestions:

- *Don't forget the hotel time.* Shoot your family waking up in the morning and getting ready. Those little encounters of people vying for bathroom time or applying sunscreen will add color, and make a nice contrast to your activity shots.

When you're on vacation, your camera bag should always hold a plastic bag and a rain poncho (see Figure 4-1). You never know when the weather is going to turn, so plan on keeping yourself and your gear dry.

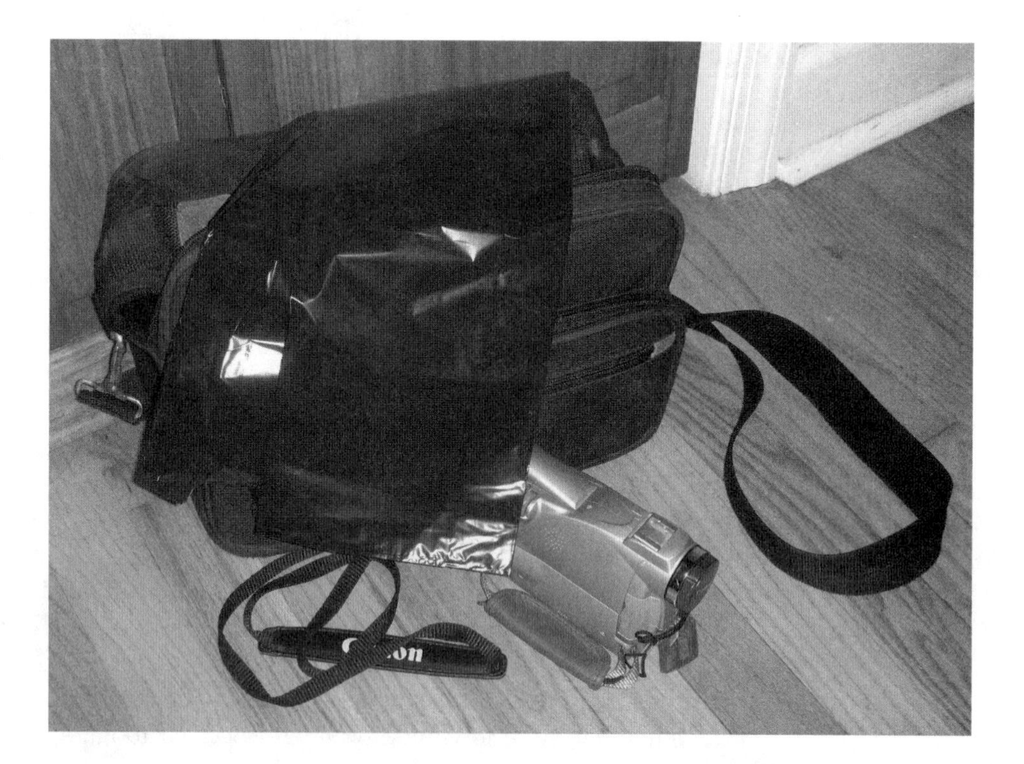

Figure 4-1

A well-stocked camera bag should definitely include rain protection.

- *If you're an early riser, take a walk with your camera before your family wakes.* If you're in a foreign town, get an early-morning shot of the town slowly waking up. If you're on a cruise ship, see what the ship looks like when most guests are asleep.

Figure 4-2

Take some shots of the countryside as you travel.

- *Get shots of the countryside flying by, as you travel by jet, bus, train, or even horse-drawn carriage.* You don't need much footage—a little is fine—but the time you spend traveling is worth remembering, too (see Figure 4-2).
- *Shoot the explanatory signs, too.* Maybe you've got great footage of Mayan ruins, but what do you do if your guests ask you about the structures and you don't remember anything? Record the explanatory signs on ruins, statues, and so on, so you can remember their significance later on.
- *Look for unusual angles.* Get up on the roof or on a balcony to shoot the plaza below, or film the boardwalk from underneath. Drop down for a child's-eye view of an amusement park. Always be on the lookout for dramatic ways to shoot—and heighten—your material.
- *Meet the locals.* A vacation movie shouldn't just be about you and your family. Talk to some of the local characters you meet along the way and ask if you can film them. You can conduct a brief interview with them, or simply record your conversation.

tip *Does your camera have a red light on the front that lets your subjects know when you're filming? Figure out how to turn it off. Look for instructions in your manual. If people don't realize that you're filming, you're going to get more candid and interesting results.*

- *Get the big and the small picture.* Imagine that you're on a busy street in a foreign city. The conventional way to record it is to walk along and film the shops and the people. The big picture shot is to get above it, perhaps from your hotel room window, and shoot the whole scene, with all the traffic and people, at once. The small picture shot is to focus on a particular person, perhaps a harried shop woman sitting on some steps and eating her lunch. Getting a sweeping shot or a detail shot is a more interesting way to paint a picture.

tip

Want to shoot in an area where your camera will get wet, such as the Irish countryside during a misty rainstorm, or on a water ride in an amusement park? Put your camera in a large plastic zip-seal bag and make a hole for the lens (see Figure 4-3). Use a rubber band to make sure the plastic stays tight around the opening. This will keep the camera electronics nice and dry. Only the lens will get wet, and you can easily dry that off afterwards.

Figure 4-3

Protect your camera from the elements and go filming in the rain.

- *Shoot your food (no, this isn't about hunting).* Food is a huge part of what makes vacations colorful and fun, but mealtimes rarely get filmed. If you're eating unusual foreign dishes, record yourself trying something new. If you're aboard a ship with vast buffets 24 hours a day, record the sight of it all.

A Shaky Camera Equals a Nauseous Audience

Your camera likely has a steady-cam feature built in to help remove some of the shake, but that helps only so much. Don't walk and shoot at the same time, and hold your camera steady while shooting. Better yet, rest it on a firm surface or buy a tripod.

If a tripod seems too large to carry, consider a monopod. A monopod is essentially a telescoping walking stick that you use to brace your camera. Because it only has one leg, it's much smaller to pack, but it still lets you take nice steady shots on your trip.

For a little more "flexibility" than what a tripod or monopod offers, consider the Gorillapod. This inventive flexible tripod lets you set your camera on rocky or uneven surfaces, places a tripod wouldn't work. Because the legs are completely flexible, you can even wrap it around poles or signposts, to get whatever shot you want. Buy it from www.joby.com. Get one of the larger sizes, unless you have a super-compact video camera.

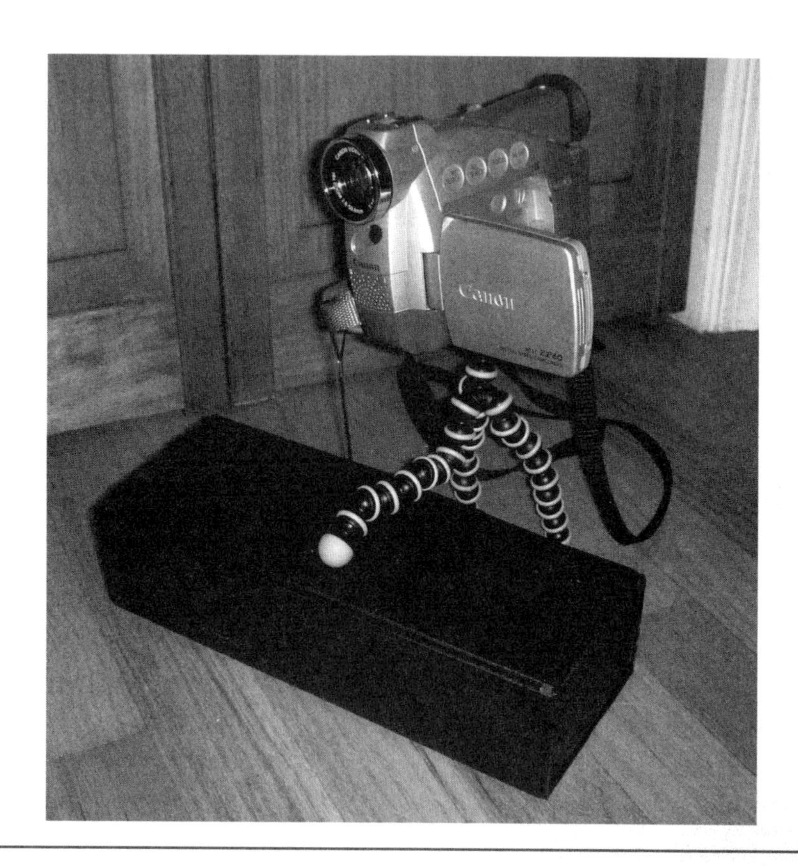

Tell a Story

While you're filming, think about the final product and how you'll want to show off your vacation. Remember, you're not just showing vacation highlights, you're telling a story. Use these techniques to help communicate your story:

- Include a few explanatory scenes, where family members tell about the next leg of the trip. Shoot someone explaining where you'll be traveling next.
- Show visual cues like maps with a route outlined (see Figure 4-4) or highway signs along the way. This will help your viewers feel like they're taking the trip with you.
- Remember that every story needs a beginning, middle, and end. Think of the arc of your story, and about how you can start, build, and end it in a satisfying way.

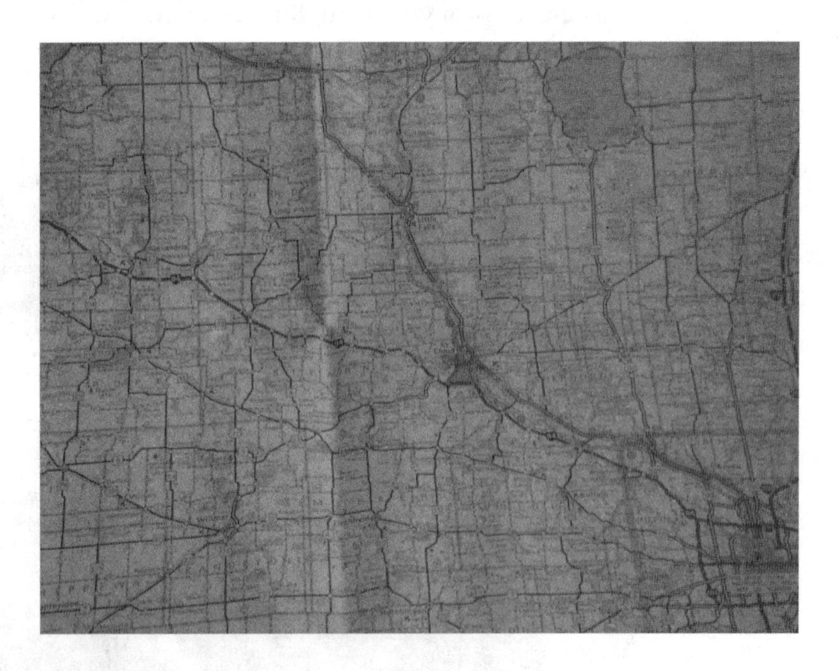

Figure 4-4

Add visual cues, like maps with highlighted routes, to help tell the story of your vacation.

tip *Record your guide, too. If you're on a guided tour, bring along a wireless lavalier microphone and ask your tour guide nicely if he or she will wear it during the tour. This will ensure that you get great audio of the tour, and it frees you to focus your camera on what's being described, not on the person doing the talking. Tip your guide well afterwards.*

Let the Kids Do Some Filming

Does the thought of sticky little fingers on your beautiful high-end video camera make you shudder? That's understandable. But wipe those fingers off and let your kids shoot a little footage. It'll make them feel like they're a part of the recording process, and the results might surprise you. Here are a few more reasons to hand over control to the budding filmmakers:

- If you let the kids do some shooting, you'll be able to get in some of the shots. Too often, the person doing the filming never gets his or her face in the movie.

tip *Study your camera's menu structure before your trip and know how to move around it. Your camera probably has settings such as Sports, Portrait, Spotlight, Sand & Snow, Low Light, and Night, which can all enhance different types of shots. But all of that functionality will go to waste if you don't know how to call up the different settings.*

- A few shots taken from a child's-eye view will look charming in your final movie. You probably won't want to leave many of them in—a little goes a long way—but having a few will give a nice change of vantage.

caution *When you're shooting outdoors, always have the sun at your back. Having the sun in front of you reduces your subjects to nothing but silhouettes. Here's a rule of thumb: if your subject looks a little dark on your camera's monitor, it'll look even worse on a computer or TV screen at home.*

Music and Memories

Pay attention to the music you hear while on your vacation. It'll make a nice souvenir of your trip, and it's great for dressing up your movie. The following are a few ways to incorporate music into your vacation movie:

- If you hear some great live music on your trip, perhaps from street corner musicians or from a small jazz combo in a tavern, record as much of it as you can onto your camera. You probably won't want to use the whole performance in your finished film, but you can separate the audio from the video track during editing, and then use that audio under your beginning titles or end credits. You could also use it to dress up an otherwise quiet scene.

Be aware of where your camera bag is at all times when you're traveling. A video camera is an easy item for a thief to grab and run with, so never let it out of your sight.

- If you're traveling in a foreign country and you hear a great song on the radio by a local artist, try to track down the CD. Go into a music store and hum a bit of it to the clerk to see if they know what it is. Not only will you get a great acoustic souvenir of your travels, but you can easily import music from a CD to make a soundtrack to your movie.

The Wrap-Up

As the vacation is ending, get a few wrapping-up shots from the family. On camera, ask each what their favorite parts were (see Figure 4-5). It's a nice way to let everyone have their say and to remember the best times of the trip.

Figure 4-5
Interview family members after the trip and ask them about their favorite parts.

The Editing Process

There are few things easier in the world than not editing your footage. Letting your used MiniDV tapes accumulate in a drawer or letting your footage simply gather unwatched on your computer's hard drive is a breeze. Far harder, for those given to procrastinating, is actually doing something with your material.

- If editing your vacation footage seems too overwhelming and you're having trouble getting started, break the work down into doable chunks. Pick one day to import the video onto your computer. On another day, split the footage into scenes, if your editing software didn't do it automatically (or didn't do a good job). You can then shape the scenes by paring down unneeded content (if a scene feels too long, it probably is; keep them short), raise and lower audio where needed, add a soundtrack, and add titles and credits. There's no reason to think that everything needs to be done all at once. If you break it down, you'll create manageable tasks and you'll be finished before you know it.
- Edit out any zooms you made while filming. Most consumer-level video cameras can't zoom smoothly, so any zooming looks herky-jerky onscreen. Remove those motion sickness–inducing zooms during the editing process.
- Shots of only five to ten seconds in length might seem too brief, but they'll keep your movie flowing at a nice pace. Try to keep the whole finished work between 10 and 15 minutes, to avoid trying your viewers' patience.
- Add a little narration during editing (see Figure 4-6). If you start putting your scenes together and realize that some parts need an explanation, connect a microphone to your computer, if you don't have one built in, and record narration over those areas. It will make the movie easier for guests to watch, and you'll appreciate the narration in years to come, when you've forgotten the details.

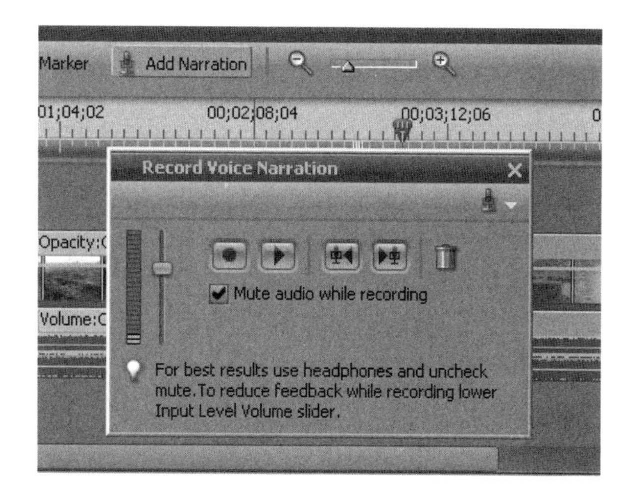

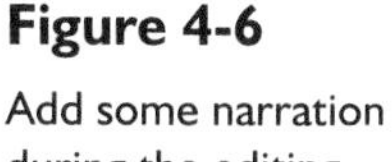

Figure 4-6

Add some narration during the editing process. This graphic shows where the narration controls are located in Adobe Premiere Elements.

- Save a few shots for a blooper reel. If a few things went haywire during your trip, like someone slipping on a wet dock or dumping a plate of food, put them on a blooper reel after the credits. Your viewers will be surprised and your movie will end with your guests laughing.

There's no step-by-step approach to creating a great vacation movie, but there are plenty of tips and techniques you can learn to highlight all the interesting things that you're seeing and doing. Study this project before your next vacation and you'll end up with a vacation movie that you're proud to show off.

Project 5
Record a Party

What You'll Need

- Digital video camera (any)
- Home computer
- Video editing and disc-burning software
- Cost: $100

A video camera might seem like the worst thing to take to a party. After all, when you're throwing a party, you want your guests to relax, be loose, have a good time—the exact opposite of what many people do when they see a camera. Yet, with a little ingenuity and purpose, a video camera can make your party a whole lot more fun and help you create movies that you and others will want to see over and over again. Use it to create video that you'll show at the party or use it in party, games that will have your guests laughing. In this project, I'll show you a variety of smaller projects that you can do to enhance and record a party.

Video for a Children's Party

It's easiest to mix video with a children's party, since children are naturally more relaxed in front of a camera. They don't have the self-consciousness of an adult and they're often thrilled to see themselves on a video screen. Try these party games the next time you're hosting a birthday bash. Not only will your young guests have a great time, but you'll get far more than the usual party footage.

The Red Carpet Interview

Watching stars arrive at an award show is fun because you get to see their glamorous fashions and see them think on their feet as they try to answer dozens of questions about who designed the clothes they're wearing and who they hope wins. Give a little glamour—and a lot of silliness—to your young guests by surprising them with your own version of Joan Rivers.

Step 1. Being Joan

You'll need a host before you do anything, so one of the adults will have to volunteer to play the interviewer. Whether it's a male or female doesn't matter, as long as you've got a blonde wig and a sharp tongue.

Step 2. Let that Light Shine

Gentle, natural lighting is usually the best choice, but for Hollywood-style camerawork, shine your on-camera light as harshly as possible to let your guests know they're being filmed. If your camera doesn't have a built-in light, you can buy an attached light for about $35 from a camera shop. This type of light fits into the accessory shoe or hot shoe on the top of the camera. A *shoe* is a U-shaped metal clip that sits on top of many digital or video cameras. A *hot shoe* can provide electricity to a flash unit, while an *accessory shoe* is just a clip. If you have an accessory shoe, good for holding a microphone or a flash in place, you'll need a battery-powered flash attachment, since it can't draw power from the camera's battery.

Step 3. Ask Away

Have the interviewer pepper your guests with questions as they arrive (see Figure 5-1), while the cameraperson records them. Be a little distracted and call people by the wrong names, just as Joan does on TV.

Figure 5-1
Interview young guests as they get to the party.

Step 4. Do a Quick Edit

After all the guests have arrived, the cameraperson should run to the computer to import the footage. (Note: there's no production team here. It's one person) Hook up your USB or FireWire connection cable and start up your editing software. Click the onscreen import option. When the footage has all been imported, quickly make a montage of each guest being asked a question. Put the timeline playhead cursor where you want your editing cuts, and then trim the material down to brief questions.

Step 5. Burn a DVD

You don't need to give a fancy treatment to this material, so simply burn it to a DVD as soon as you're done. Click the onscreen option to export the footage to a DVD. In Ulead VideoStudio, for example, select the Share tab at the top of the screen and then choose the Create Disc option on the left. This opens the DVD-creation wizard, shown in Figure 5-2.

Figure 5-2

Burn a quick DVD in Ulead VideoStudio.

Step 6. Critique the Fashions

Later in the party, have your Joan Rivers impersonator return. Organize your young guests around the television and play the DVD. Stop the movie on each person and let your Joan impersonator critique what that person is wearing. Be sure to have a harsh word for everyone, but in an over-the-top silly way that won't offend. Finally, present awards such as Best Dressed, Loudest Outfit, or Most Likely to Appear on MTV—whatever works with the outfits. Make sure everyone gets a prize.

The Music Video Party

Young girls especially like to sing karaoke, acting out scenes from favorite videos. Use that to have them create their own entertainment for a birthday party. They do all the work, while you get to hang back and film the whole crazy scene.

Step 1. Setting Up

If you're going to give partygoers a stage to sing on, you need to do some preparation. Pick up a few colored lights from a well-stocked hardware store. Or be creative with strings of Christmas lights. If you have access to a glitter ball, you've got it made. Prop up a flashlight or two to shine on it while your guests sing.

You'll also need something for your guests to sing to. If you already have a karaoke machine, you're set. Many stereos have a karaoke mode that strips out the vocal tracks of whatever is playing—not perfectly, but good enough to sing over. You could also consider iKaraoke from Griffin ($49.99; www.griffintechnology.com), a clever microphone and processor all-in-one. It attaches to any iPod, as shown in Figure 5-3, and reduces the volume of the lead vocal, plus it even lets you add reverb.

Figure 5-3
The Griffin iKaraoke turns your iPod into a karaoke machine.

Step 2. Camera, Meet TV

This isn't just a karaoke party but a music video party, and music videos are watched on the television. To add an MTV feel, you need to connect your camera to your television, so that your guests can watch the action live.

Your video camera likely came with an analog RCA cable for carrying composite video and stereo audio. If so, you'll have a cable with three differently colored plugs on one end and a single plug on the other (see Figure 5-4). The single plug goes into the AV port on your camera. The AV port is likely yellow, and hidden underneath a flap somewhere on the front or side of the camera. The three plugs on the other end will be red, white, and yellow. The yellow plug carries the video signal, while the red and white plugs carry two channels of audio. If this is what you have, connect the three plugs to the matching colored ports on the back of your television.

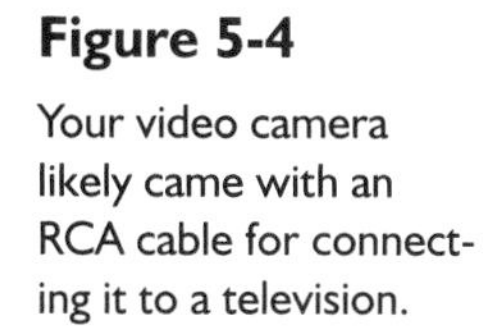

Figure 5-4

Your video camera likely came with an RCA cable for connecting it to a television.

If your camera supports widescreen display, and you'd rather show a widescreen image, you can connect an S-video cable to your camera and the television for higher-quality video. In this case, you still connect the red and white plugs from the RCA cable, since those deliver the stereo audio, but you don't connect the yellow video plug. Instead, the S-video cable delivers the video.

tip *If you're going to show the material on a widescreen set, make sure you're shooting in widescreen, to take advantage of all that space. Call up your on-camera menu and change the size to widescreen or 16:9 (the ratio for widescreen video).*

The connections previously mentioned are all analog connections found in lower-priced cameras. If you have a higher-end TV set and video camera, you'll want to connect them with a digital High-Definition Multimedia Interface (HDMI) cable. If your camera supports HDMI but didn't come with a cable, you'll need to buy it separately.

caution *Analog cables, like RCA cables, can vary in quality, but digital cables, such as HDMI cables, are all the same. There's no point in spending more than $20 for an HDMI cable. Also, don't be confused by version numbers attached to HDMI cables. Those refer to the devices that use them. The cables themselves are all always the same.*

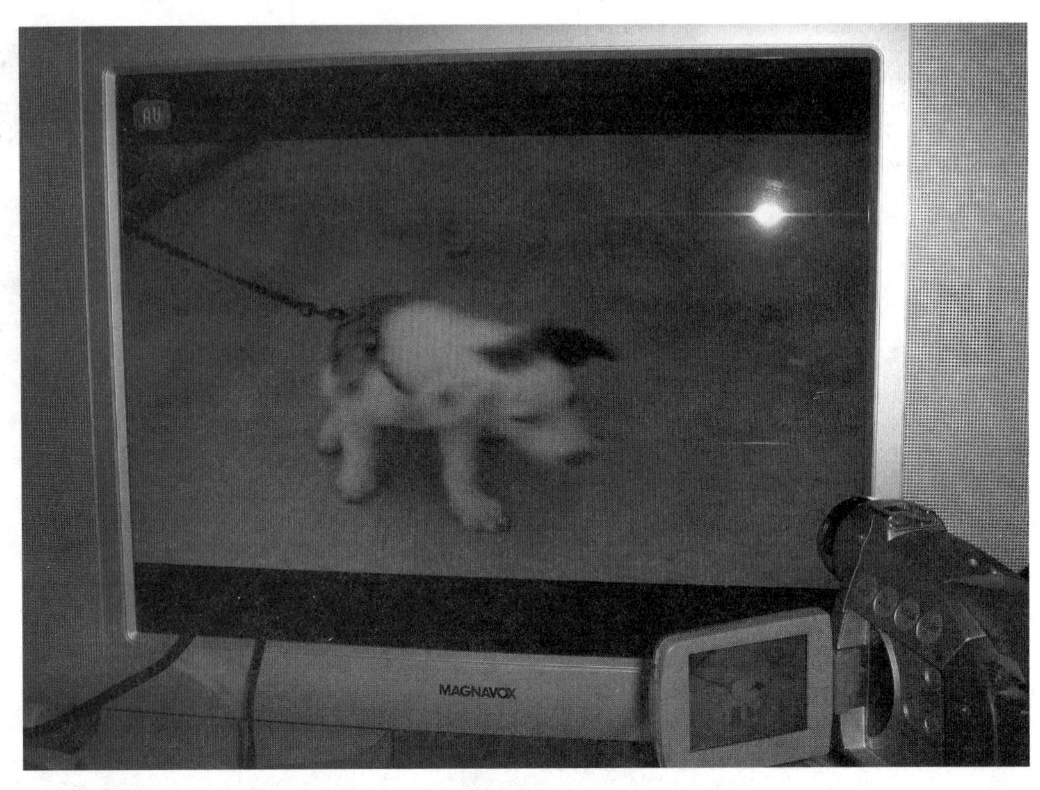

Figure 5-5
You'll need to adjust the source for your TV to input from your video camera.

Adjust the source setting on your television, until you see the camera viewing screen (see Figure 5-5).

When your video camera and TV are connected, you can use the television as an external monitor for viewing what's on the video screen. Put your stage area in front of the set and slightly to one side, so that the other guests can see the singers and the TV image at the same time.

note *These instructions for connecting your video camera and television are useful anytime you want to show people what you've recorded. Video cameras have both recording and playback modes. If your video camera is in the recording mode, the TV acts as an external monitor. If it's in playback or VCR mode, you can use the TV to show footage you've recorded. This is useful for times when you want to show someone material you've recorded, but you don't have the time to edit it or burn it to a DVD. Simply connect the correct cable or cables and you're ready to go.*

Create Digital Party Favors

Another way you can use a video camera at a children's party is to create a favor or souvenir for the guests to take home. If you're able to work quickly—and have a small guest list—you can burn some DVDs while your guests are finishing their cake.

Step 1. Host a Contest

Make the main entertainment a contest of some sort, such as a skit contest or a dance contents. Divide the guests into teams and give them time to work out a skit or a dance routine. Provide some sort of costumes or props to jog their creativity.

Step 2. Record the Results

Set up your tripod to record the whole of the contest. If you're using low lighting, remember to adjust the light setting on your camera. To do this, call up the menu and look for lighting or filter settings. Choose the low-light setting (see Figure 5-6) and then switch back to record mode.

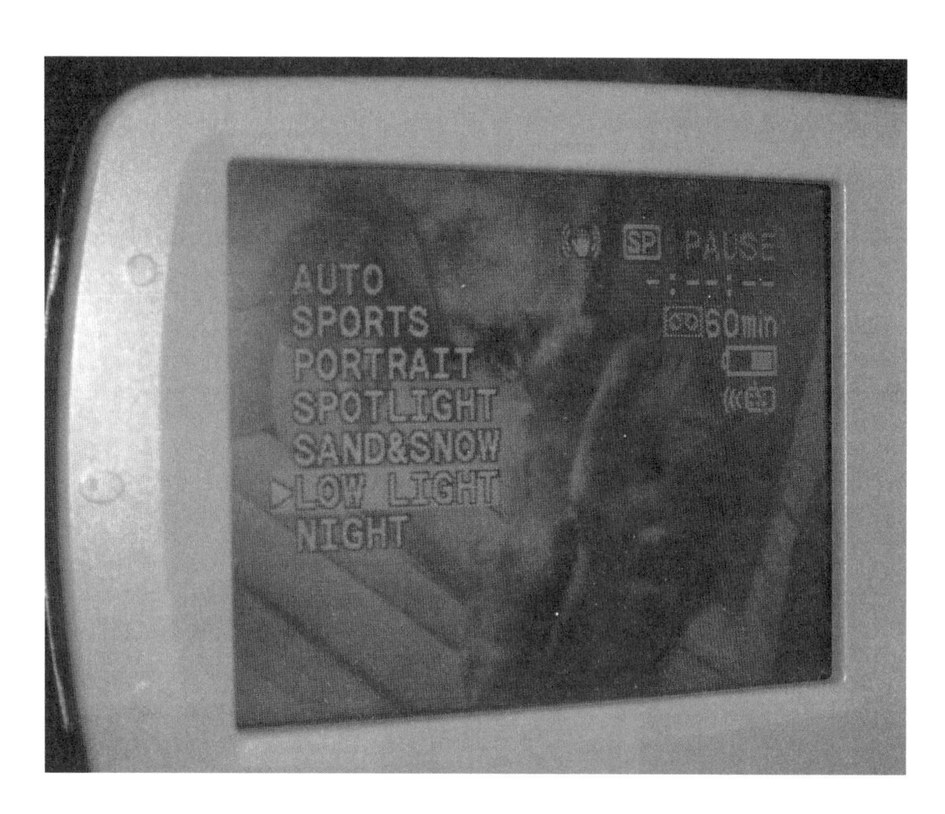

Figure 5-6

Adjust the light settings if you're shooting in a dimly lit environment.

note *Ever wonder how well a particular video camera can perform in low-light conditions? A camera's ability to shoot in low light is given in "lux." Lux is a measure of light intensity. To give you some idea, moonlight is under one lux, while a candle one foot away is ten lux. The lower the number of lux that your camera can work in, the better it will be at low-light shots.*

Step 3. Do a Quick Burn

When the contest is over, quickly take the camera to your computer and input the footage. Call up your video editing software, connect your camera with your USB or FireWire cable, and select the import option. You won't have time for much—if any—editing, but you should add a title saying whom the party is for and giving the date.

Select the onscreen option to burn a DVD. Because your footage is short, this shouldn't take much time. As soon as one disc is done, put a fresh blank disc in your computer and burn another. As long as you keep your DVD software open, you'll be able to burn identical discs without touching your settings.

When your discs are finished, load them into CD cases that you've prepared ahead of time. Many consumer-level DVD-burning programs come with CD label-making software built in.

tip *If you're using iMovie and iDVD on a Macintosh computer, you don't have label-making software included. You can find inexpensive shareware programs online, but consider buying Toast (see Figure 5-7), a disc-creation suite from Roxio ($99; www.roxio.com). It includes labeling software and much more, including tools to back up, store, and archive your data, and to create many different types of CDs and DVDs.*

You're then ready to distribute the burned DVDs as party favors before your guests go home, letting them remember the fun as soon as they get home.

Another option is to immediately upload your party footage onto YouTube or another hosted video site, and let your guests know before they leave where they can

Figure 5-7 Roxio Toast's disc-labeling software can help Mac owners create DVD holders ahead of time.

find it online. They might even prefer this, because if the video is online, they can post comments and read other people's comments. For detailed instructions on posting to YouTube, see Project 22.

tip *If your DVD-creation software supports it, you can burn video to CDs in the form of Video CDs (VCDs) and Super Video CDs (SVCDs). These types of discs will play in all home computers and in most DVD players. They offer less storage space and lower video quality than do DVDs, but you get the advantage of using less-expensive discs. It's also a good option for people who don't have a DVD burner.*

Video for Grown-Up Parties

An adults' party doesn't offer the same opportunity for silly fun as a kids' party, but you can still use a video camera in different ways.

The Anniversary

If you know of a major anniversary coming up, or even a major birthday, a video testimonial is a great way to mark the occasion.

Step 1. Conduct Interviews

You need to give yourself plenty of time for this one, since you need to interview several people close to the guest or guests of honor. Set up your tripod in each person's house, for a good steady shot, and ask questions.

tip *Professional camera operators swear by the rule of thirds, and you should, too. Imagine a tic-tac-toe grid over the screen. Try to keep the subject or subject along one or more of the lines. It creates a more pleasing composition. So, for example, if you're interviewing one person, instead of putting his or her face in the center of the screen, put their image where one of the gridlines would be.*

Step 2. Edit the Footage

You want your end movie to be tight and not too long. Use titles to introduce questions that you ask more than one subject. For example, put a title onscreen that says "How long have you known Alice and Sam?" and then show the responses of several people to that question. In this way, you don't need to show yourself asking the question over and over. Title options are available as one of the main options in your editing program. Keep titles a few seconds long, so everyone has a chance to read them.

With a little planning, your video camera can add flare to your party and help you capture memories. Charge up your camera before your next get-together and put it to work while everyone else plays.

Project 6

Make a Music Video

What You'll Need

- **Digital video camera (any)**
- **Home computer**
- **Video editing and disc-burning software**
- **Cost: $100**

Whether you have a serious musician in your house or just a few lip-synchers who like to have fun, making a music video is a great project for the family. If you're in a band, you can make a music video to show off your latest single. Post it to YouTube or a MySpace account to show the world what you can do and to win more fans. If you're applying to a college as a music major, you can create a music video and burn it to a DVD as part of your application. Or, if you and some friends simply want to have some fun and be creative, you can shoot a video to someone else's material, and then post it to a site like YouTube so that others can rate and share your work. (For more info on posting to YouTube, see Project 22. Posting a video with someone else's music should be considered fair use, as long as you're not profiting by it.)

Creating a music video can be as involved or as casual as you like. You can shoot it all in one take or spend a lot of time planning the look. In this project, you'll learn all the steps of shooting and editing a good music video at home. Add your own creativity for amazing results.

Step 1: Planning Your Video

If you're just out to goof off in front of the camera and have a little fun, you can skip this step, but if you want to make a video people will watch, read on. Even if you only plan to post your work online, you should do some planning. You won't get a top-viewed video on YouTube without planning (well, unless you're spectacularly bad, that is, but that can't be taught).

Storyboarding

Storyboarding means diagramming shot by shot what you plan to record. The next project goes into detail about how to storyboard. It's an advanced step, but a necessity if you want professional results.

If full storyboarding is too much work for you, consider a compromise, a light version that will let you plan your shots:

1. Type or paste the lyrics to your song into a text document. Double- or triple-space the document and then print it out.
2. Draw vertical lines to indicate where you'll change the camera angle. Break up the lyrics into sections, based on the different camera shots. Above each section of lyrics, write the type of shot you'll record. You might start with a long shot that shows the entire band, move in for a close-up on the lead singer, and then change to a location shot filmed somewhere else.

Think of a Theme

A good video isn't just bunch of musicians picking up their instruments and playing; a good video has a theme. Think of a visual theme to add to your shots, to add visual interest, build upon the song's meaning, and tie all your shots together. Here are a few suggestions:

- Think of a visual element that you can repeat in your shots. You could have the musicians surrounded by loads of candles (see Figure 6-1), or have lots of streamers hanging from the ceiling. Throw a strange object such as a garden gnome into every shot or have everyone wear the same color.
- The location can be the theme. You could shoot the video in an unusual space, such as a tree house or a city bus. Choose a place that ties into the lyrics of the song and creates a deeper meaning.
- You don't even need to use people in your video; you could let toys or props tell your story. Stop-motion effects animate still objects and are a lot of fun for viewers to watch. For details on creating a stop-motion movie, look to Project 17.

Costuming

If you're filming this video for a serious purpose, like a college application, wear your best suit. But if you're having fun and playing rock star, you'd better dress the part. Isn't putting across a rock star attitude part of the fun of shooting a video? Play it up; the more striking the costume, the better.

Figure 6-1

Think of a visual theme for your video.

Step 2: Choose Camera Effects

Your camera probably has some built-in shooting effects that would enhance a music video. These instructions tell how to access them for some cameras, but the exact steps might not match up with the video camera you're using.

Fade-In/Fade-Out

To create automatic fades when you start or stop shooting, follow these steps:

1. Change your camera's setting from the easy mode to the program mode. By default, some cameras are in easy mode so that you can simply turn them on and shoot, getting good results with auto settings. If you want to change advanced settings, you need to change the mode. Look for a switch near the LCD screen that lets you change modes. If you can't find one, consult your manual.

2. Turn your video camera on. Activating an effect can be tricky on some cameras. Press the Digital Effects button and you'll see an onscreen menu of available effects. Scroll to the one you want. You might think that clicking your choice will activate it, but that isn't so. Instead, press the Digital Effects On/Off button on the camera itself. This selects the particular effect, but you still need to activate it whenever you want to use it.

3. You'll now see the effect name flashing on the screen. To use it, press the Digital Effects On/Off button one more time (see Figure 6-2). Now the fade effect is active. Click the Record button and you'll see your shot slowly fade in from a black background.

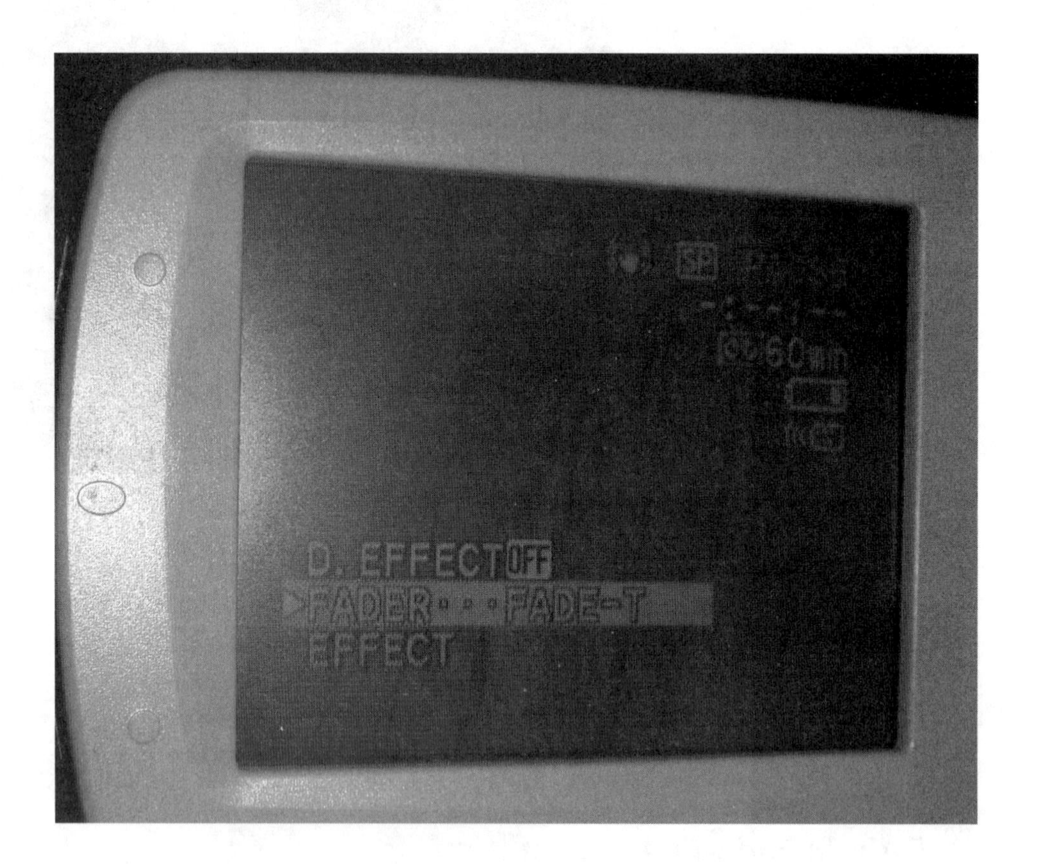

Figure 6-2
Set your camera for automatic fade-ins and fade-outs.

tip *Besides the fade-in/fade-out effect discussed here, many cameras have wipe effects in which your image starts as a vertical or horizontal sliver and expands to fit the whole screen. Experiment to see which effects you want to use.*

4. The fade was only active for that one shot, for the fade-in. If you want to fade out at the end of your shot, you need to activate it again. You should see the Fade name flashing on your screen again. Click the On/Off button while shooting to activate the fade again. Now when you stop recording, your footage will slowly fade to black.

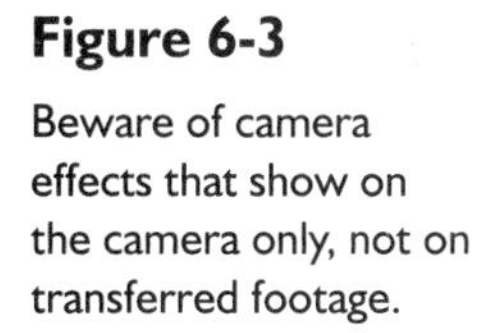

Figure 6-3

Beware of camera effects that show on the camera only, not on transferred footage.

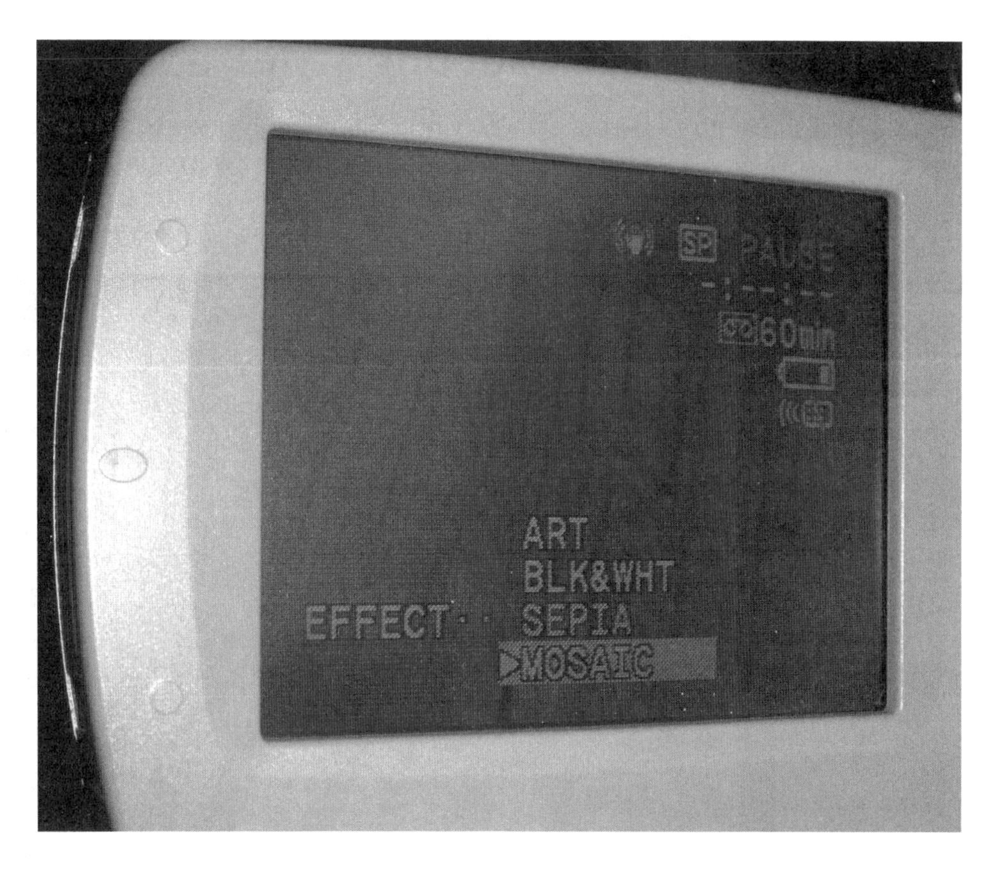

caution *Beware of effects that show back only on the camera itself (see Figure 6-3). Some cameras, for example, offer a mosaic fade, in which your footage starts as a heavily-pixilated mosaic image and then slowly fades into sharp focus. But this effect isn't transferred to the footage itself; it's done on-camera and only shows when you view your footage in the VCR/Playback mode.*

Other Camera Effects

If you're going to edit your footage with a software program, you have the option of adding digital effects during the editing stage. This is the preferred way to go, because it gives you the freedom to change your mind at a later stage and remove the effects. However, your camera also has built-in effects you can use if you don't have time for editing. These likely include black and white, sepia, and mosaic effects.

Choosing an effect is much like choosing a fade:

1. Press the Digital Effects button and you'll see a list of possible effects and fades. Scroll to the effect you want, such as the mosaic effect.
2. Select the effect you want by pressing the Digital Effects On/Off button. You'll now see the effect name flash on your screen.
3. To activate the effect, press the Digital Effects On/Off button one more time. Now you should see your screen change to reflect the effect you've chosen. Unlike with fades, you need to select an effect only once to use it for a whole recording session.

Step 3: Decide on Shooting Styles

Say no to boring camera work. Shooting a video is as demanding as home video gets. The videos on television often switch between many different camera shots, and yours should, too (see Figure 6-4).

Figure 6-4

Get a range of shots for your video.

Be sure to use these types of shots, if you're filming a band playing (or pretending to play):

- A long shot showing the lead singer with the band arranged behind him or her.

Want to get a great rocker performance out of your lead singer? Then set up a mirror behind the camera. The singer will get instant feedback about how he or she looks, and will have confidence to really perform for the camera.

- Medium shots showing only one or two of the band members.
- Close-ups on each of the band members.
- Odd-angle shots: put the camera down on the floor and shoot up, or rig the camera so that it's directly over the band looking down.
- A pan shot. Move around the band members as they're performing. Do this as smoothly as possible, to avoid giving your viewers motion sickness.

tip *If the band members are lip-synching, you can yell out cues to them while they're performing, because you're not going to use the audio track from the footage anyway. You'll delete it during editing and add in a clean digital track from a CD or from a prior audio recording session. That frees you to yell out directions and cues during the actual filming.*

Step 4: Edit Your Video

If you're going to make a fast-paced video, roll up your sleeves and get to work. They don't call quick cuts "MTV-style editing" for nothing.

The Automatic Cheat

The secret to making a good video is timing your cuts to the beat. Some programs have video-creation wizards (see Figure 6-5) that do this automatically and turn out surprisingly sophisticated videos. To use them, follow these steps:

1. If you're using Pinnacle Studio, start up the program and select the SmartMovie option.
2. Connect your camera to your computer with your USB or FireWire cable and import your footage. If it's already loaded onto your computer, simply import it into the program.
3. Move to the Edit tab at the top of the screen. Click the movie camera icon midway down on the left side. You'll get a list of video styles to choose from. Pick the one that suits your music.
4. After you've picked a style, enter a title and some end credits.
5. Close the selection area so that you can see your video clip library again. Drag the clip or clips you want to use down into your clip area. Click the Return to SmartMovie option above the clip area when you're done.

Figure 6-5

Some editing programs, like Pinnacle Studio, have wizards that create music videos automatically.

6. From the selection screen, choose a song to use. You can take one from your hard drive or import one from a CD.
7. When you're done, click the Create SmartMovie button. The software will create a video to your song with edits cut to the beat.

Standard Editing

If you want more control over the finished product, you'll want to do the editing yourself. Here are the basic steps to follow:

1. Import your footage using your USB or FireWire cable.
2. Unless you want the live sound from your shoot, you need to delete the audio track. In Apple iMovie, for example, drag your footage down to the timeline area and then click your footage once to select it. Choose Advanced | Extract Audio (see Figure 6-6). The audio extraction can take a few minutes. When it's done, you'll see the audio information for that clip loaded into the first audio track of the timeline.

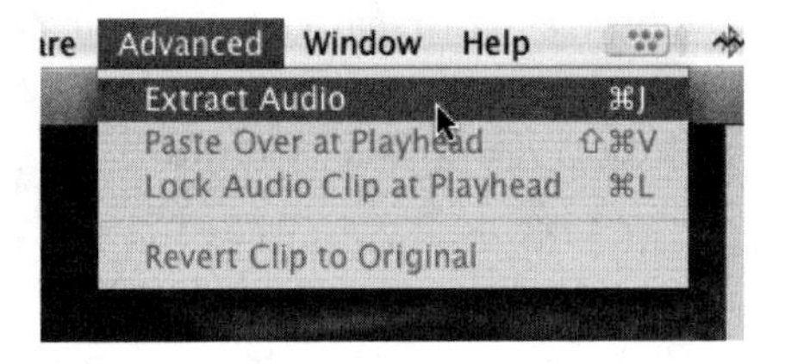

Figure 6-6

Apple iMovie makes it easy to extract audio from your footage.

3. Click the audio track once to select it and then press the DELETE key. Your audio will vanish in a puff of onscreen smoke.
4. Now you're free to add your own music. Import the song you'll be using from a CD or from your hard drive.
5. Trim your shots down so that you have the best take for each line of the song. Look to your storyboard or shooting script to jog your memory about what footage you shot for which lines of the song.
6. Time your edits to the beat of the music. To heighten interest and make your work look professional, change the shot on heavily accented beats.

Onscreen Effects

Every editing program comes with a list of onscreen effects that you can use to add atmosphere to your footage. To see effects in Ulead VideoStudio, for example, select the VideoStudio Editor option at startup and load and edit your video as normal. Click the Effect tab at the top of the screen, and then click the pop-up menu near the top left to select the Video Filter option. You'll see the various filters that the product offers (see Figure 6-7), such as:

- **Blur and Average** Blur adds a little fuzziness to your image; Average adds a lot.
- **Bubble** Shining bubbles float on your screen and even reflect the action going on around them.
- **Cloud** Misty clouds float across your screen.
- **Kaleidoscope** Your video looks as though it were seen through a kaleido-scope.
- **Old Film** Give an old-time newsreel feeling to your movie.
- **Rain** Have your singers perform in the rain, without any of the equipment getting wet.

It might sound complicated, but creating a great music video is just a matter of knowing a few techniques. Have fun with it: get out your video camera and be the rock god (or rap god or country god) that you've always wanted to be.

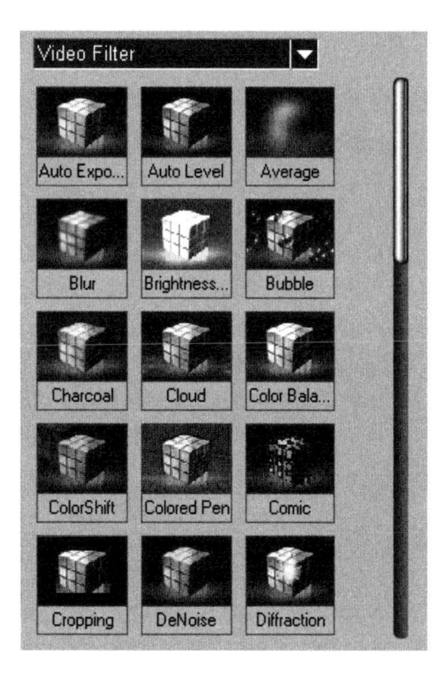

Figure 6-7

All consumer editing programs include effects you can add to your video, such as these from Ulead VideoStudio.

Project 7

Shoot Your Own Script

What You'll Need

- **Digital video camera (any)**
- **Home computer**
- **Cost: $0**

If you enjoy creating scenes and funny moments in front of your camera, there's going to come a day when you want to shoot your own script. Maybe you'll want to create a two-minute skit for YouTube. Maybe you'll want to record a ten-minute scene to entertain your friends. Or maybe you'll want to test the waters for a possible film career and shoot a full-length movie. In any case, knowing the standard formatting for your script will save you a lot of time and get you the results you want.

This project will take you to a new level of professionalism in your video and it's essential for anyone who wants to really create with his/her camera. If you've got an idea that you want to capture on video, you'll be happier with the results if you give your fellow performers written dialogue instead of asking them to improvise. And if you have ideas for scene directions, effects, or camera angles, get those down on paper while you're feeling creative. Then you can shoot your own original script, and know the pleasure of being a real director.

The Perfect Script

You could type out your script in any old format, but the industry-standard shooting script has evolved for a reason. It lets you and your actors see in a clear and simple way not only the dialogue, but also the onstage action and the camera directions. Follow these steps to create a perfect Hollywood shooting script.

Step 1: Write a Slug

Begin your scenes with what's called a "slug line." This is a simple establishing direction in the form "INT. HANSON FAMILY KITCHEN. DAY." Use the form of putting INT or EXT first, for interior or exterior, then the place, and then DAY or NIGHT.

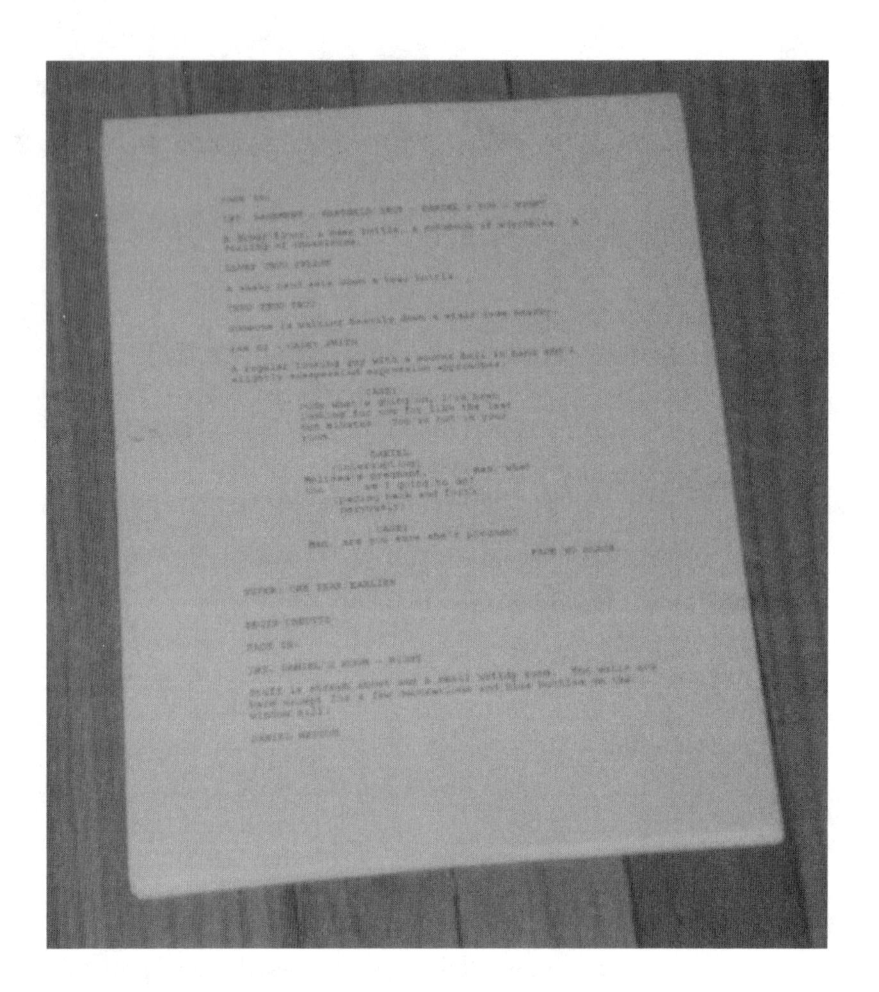

Figure 7-1

A typical film script

Don't bother putting specific times of day. Putting the slug line in all capital letters lets people reading the script quickly get their bearings. The slug line is flush left, meaning it's aligned to the left side of the page (see Figure 7-1).

Step 2: Give Direction

Underneath the slug line, put your first lines of camera direction and stage business. Camera directions go in all capital letters, while the rest of the sentence gets standard capitalization, such as "CLOSE-UP on teapot, which is boiling and whistling. CAMERA PULLS BACK: We see KRYSTA HANSON turn off the flame and pour water into a mug." Note that you also use all capital letters the first time a character is introduced. If you prefer, you can put camera directions and stage directions on separate lines, to make them easier to read. Stage and camera directions are also aligned with the left edge of the page.

tip *Be specific with your action directions. Spell out exactly what you envision. Don't just say that a character is angry; show him kicking a trash can or balling his hands into fists.*

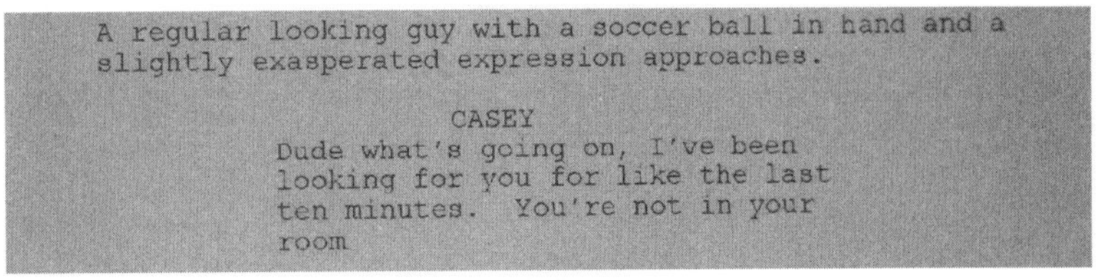

Figure 7-2

Dialogue is centered in a movie script.

```
A regular looking guy with a soccer ball in hand and a
slightly exasperated expression approaches.

                    CASEY
          Dude what's going on, I've been
          looking for you for like the last
          ten minutes.  You're not in your
          room
```

Step 3: Add Dialogue

If you've ever read a play, you'll notice that presentation of dialogue is different in a shooting script. Here, you add dialogue by first writing the character's name in all capital letters, centered in the middle of the line. Immediately under that, you put the line of dialogue, also centered, as shown in Figure 7-2. Note that sections of the script should be separated by double-spacing, but each element itself is single-spaced. So you might have a camera direction, then a double-space, followed by a long single-spaced description of some action, followed by a double-space, followed by single-spaced dialogue.

Step 4: ...and Repeat

Repeat the preceding three elements. Remember that every new scene, no matter how brief, gets its own establishing slug line.

You'll need to add a lot of camera angles to your shooting script, so master these types and terms before you start:

- **Close-up, CU, C/U** A close shot showing only a part of a person or an object.
- **Medium shot, MS** A medium-view shot, showing a few people and more of the background.
- **Long shot, wide shot, LS** The camera is pulled back to show all of the scene.
- **Extreme close-up, ECU** This focuses intensely on one small detail.
- **Extreme long shot, ELU** The camera is pulled back further than it is for a long shot, perhaps to show a whole city or town.
- **Follow shot, pan** The camera follows the action, or pans across the area.

tip *Adding dolly shots, or tracking shots, is a simple way to make your video look more professional. In a dolly shot, the camera is on wheels and slowly moves during the scene to follow the action or reveal information. Professional crews pay hundreds of dollars for complex dolly operations, but you can make a working one with an old baby carriage. Put a board across the top of the carriage, and then set your camera on that flat surface. Turn your camera on at the start of the scene and gently push the stroller as you're recording. A rolling chair will also work (see Figure 7-3).*

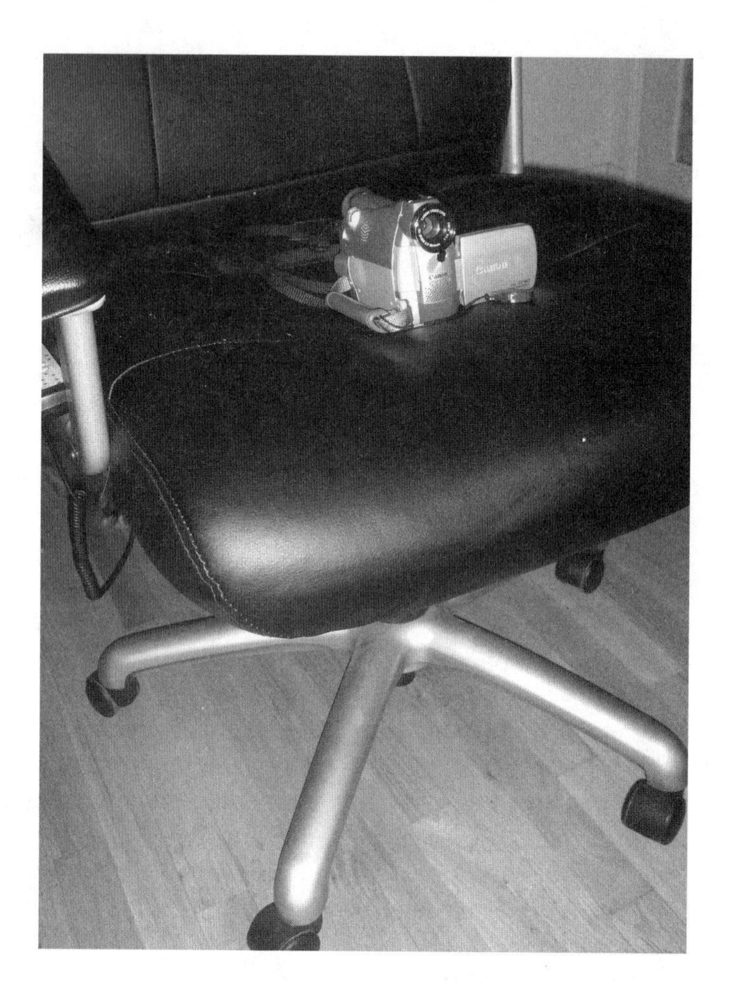

Figure 7-3

Even a rolling chair can help you create a dolly shot.

While not camera terms, you might also want to use these abbreviations in your script:

- **VO** Voiceover
- **SFX** Special effects
- **BG** Background
- **FG** Foreground
- **POV** Point of view
- **OTS** Over the shoulder

tip *Want to see how the pros write scripts? Plenty of shooting scripts for famous movies are available online for free download. Do a web search to find them. Reading through a few will help you get a feel for script conventions before you start to write.*

Storyboarding

Think of storyboards as a visual script for your movie. You've probably seen storyboards in behind-the-scenes shows on the making of a movie. They're a way to communicate the action of scenes so that everyone involved knows what's going to happen. They're also a way for you, the director, to work out how shots should move.

tip *Understand the purpose of different shots and use a mix of them. Some shots only create atmosphere, some show action, and some show dialogue. Consider all three types when planning your shots.*

Don't worry if you can't draw. Storyboarding isn't about making great finished art. Using stick figures and arrows is fine. In fact, it's pretty common. Plus, creating the drawings quickly helps you to maintain an energy while you're creating them and to avoid getting bogged down in detail. Work fast and loose as you sketch out your storyboards (see Figure 7-4).

tip *Don't storyboard your entire movie, unless it's a short one. Storyboards are typically used for action sequences, when you are using several different shots in a short period of time and the director needs to make clear exactly what all those shots are. For slower scenes, you can get by using a standard script, as described earlier in this project, or a video script, as described later.*

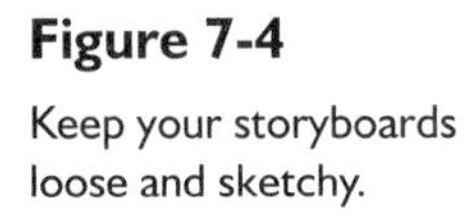

Figure 7-4
Keep your storyboards loose and sketchy.

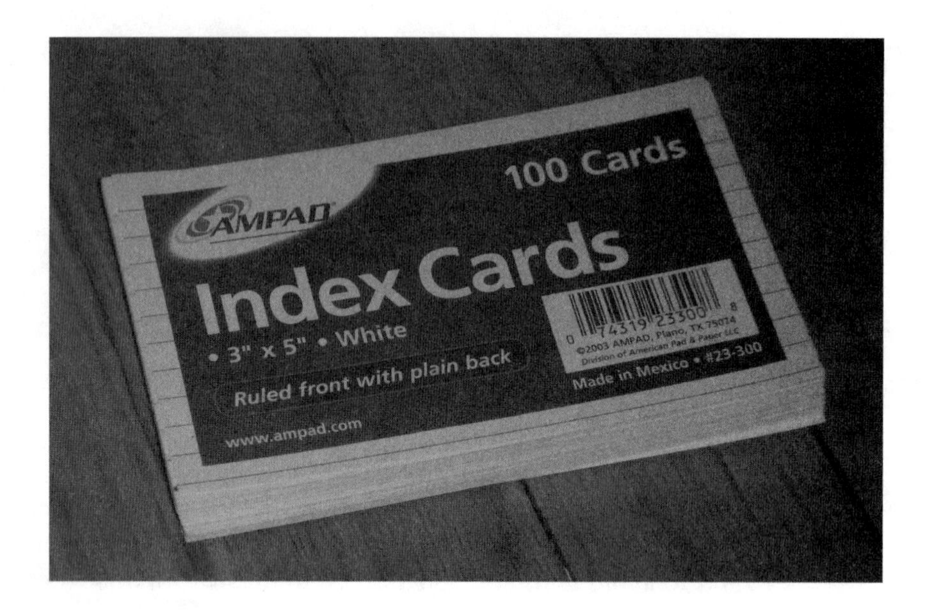

Figure 7-5
Use blank index cards to create your storyboards.

Follow these steps to make great storyboards for your script.

Step 1: Create Some Boxes

Either sketch out boxes on a large sheet of paper or buy a few stacks of large index cards, such as the type shown in Figure 7-5. Index cards work better, because you can simply toss them if you make a mistake, and you can easily rearrange them if you decide to change the shooting order.

tip *Create storyboards in the ratio that you're filming, to understand the exact dimensions you'll be working with. That will help you plan the visual space correctly. Standard video is in the 4:3 ratio, meaning you should have four units across and three units down. If you're shooting widescreen, the ratio is 16:9.*

Step 2: Start Sketching

Sketch out the action of your scenes with stick figures and arrows. If you don't draw well, use just enough visual detail to enable people to see what's going on. Don't forget to sketch out the backgrounds, as well. You need to think about the full visual space, so sketch the world that your characters inhabit.

caution *Storyboard in pencil and keep an eraser nearby.*

Step 3: Point the Way

Use arrows to show movement within a scene and to show the movement of the camera (see Figure 7-6).

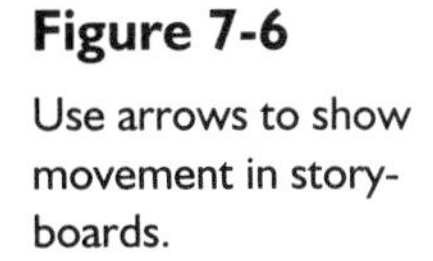

Figure 7-6

Use arrows to show movement in storyboards.

Step 4: Major Movement

If you need to describe complex camerawork or complex movement within a scene, use two or even three storyboards to do so. Label them A, B, and C so that people understand they're all parts of the same shot.

caution *Camera movements should have a point. Don't set up a dolly shot just because you can or because you think the scene would look too static otherwise. Camera movements need to have a purpose, showing or revealing something to the viewer. Think of each new camera shot as a tool to convey your story to the viewer, and choose shots that are effective in telling that story.*

Step 5: Zoom In

Use boxes within your storyboards to convey zooms. For example, if you're starting out with a medium shot on a character as she considers making a telephone call, and then plan to zoom in to show only the phone, sketch the whole medium shot with the character and the background. Next, draw a box around the area that you'll zoom into, in this case the telephone. Use arrows at each of the four corners of the inner box to show how the larger shot will zoom into the tighter shot.

tip *If you're more comfortable with a keyboard than a pen, use a design program to create your storyboards.*

Step 6: Line Up Your Storyboards

When you've created all of your storyboards and are happy with the order, number them. This will help you refer to shots quickly, so you and your crew know which one you're working on.

note *The more storyboarding you do, the less shooting you'll have to do. Storyboarding lets you work out shots on paper before you assemble your crew, so that when you're ready to shoot, you'll know exactly what you want.*

You'll need to give a little textual information with each storyboard as well, giving camera directions and scene directions, but how you place it is up to you. Here are some options:

- You can add two text boxes below each image, the top one giving camera directions and the lower one providing scene direction (the action that the characters in the shot will perform). So, the top box tells what type of shot each storyboard is and describes how the camera needs to move, while the lower box tells what actions the actors need to make during that shot.

tip *Don't add text to the storyboard image itself. You're not creating a cartoon. Put text either above or below the image, so that the diagram itself is kept purely visual.*

- If you find all that description unnecessary, you can just write a brief camera direction under each image. This makes it easier to take in all the information from a storyboard at a glance.

tip *If you're filming and you get a great idea that isn't on your storyboards, run with it. Your storyboards are a guide, but they're not handcuffs. Feel free to deviate from them if it helps the film.*

The Video Script

Somewhere between the movie script and storyboards lies the video script. This is a text script with two columns, one for camera directions and one for the dialogue and music cues (see Figure 7-7). It offers a quick visual way to see what happens in every camera shot.

You can use just about any document program to create a video script. I'll give directions for the most popular one, Microsoft Word.

Step 1: Create Your Script Framework

1. Open Microsoft Word and choose Table | Insert | Table.
2. Create a table with one row and two columns. It won't look like much now, but you can build on it as you go.
3. Put the title VIDEO above the first column and the title AUDIO above the second.

Figure 7-7

The video script

The Underage Detectives

Video	Audio
Exterior shot of house, pan over the lawn and to the front door.	Music Under: Somewhere Over the Rainbow Narrator: For a while when I was young a family lived down the block from us, a family that was completely unlike my own—and I was fascinated by them.
Close-up of toaster as two slices of toast pop up	Music: Big band music on radio Dad: Eat up fast. We're already running late.

Step 2: Add Directions and Script

Start adding your directions and script. If you've already typed your script, transfer the dialogue for one shot and one shot only to the table cell on the right. Use the standard stage play format for dialogue, with the name of the character speaking, a colon, and the dialogue all on the same line (refer to Figure 7-7). All audio goes in this cell, so include any music cues.

Step 3: Set the Shots

After you've inserted the dialogue, add camera directions to the left cell (refer again to Figure 7-7). Put camera directions next to the line of dialogue that they tie into. So, for example, if the camera needs to zoom in on a character's face when the character says a certain dramatic line, put the zoom cue next to that line.

Step 4: Grow Your Table

Use a new table row for every shot. To easily add a row just like the one you have, place your cursor to the right of the row, so that it's just outside your table, and press ENTER or RETURN. You'll get a new table row underneath with the same number of columns.

Step 5: Making a List

When you're done, number your scenes. This will help you quickly refer to them when talking to your crew.

As you've seen, creating a script isn't too much work, and your videos will look a lot more professional and planned out if you do so. Writing out a script will ensure that you create the kind of movie you're dreaming of, and knowing how to create the right kinds of scripts and storyboards will ensure that you get the shots you want.

Project 8

Create a Video Family Album

What You'll Need

- **Digital video camera (any)**
- **Home computer**
- **Movie editing software**
- **Cost: $100**

Using a video camera is all about making memories, right? Recording your family's special moments through the years? Well, instead of having one video for one occasion and another video for another, why not create one special DVD that has all of your relatives in it? Think of it as a video family album.

In this project, you'll be updating the traditional family photo album for the digital age. Instead of including photos of all of your family members, you'll include interviews or conversations with them. Nothing stays the same, and families grow, change, and even lose members all too quickly. But with this project, you'll be making a living recording of your family, exactly as it is right now. Out of all the projects in this book, this is the one you'll cherish the most as you get older. Record some of grandma and grandpa's stories while you're still lucky enough to have them around. Record your great aunt whipping up her famous cookies. Get footage of the baby cooing over a stuffed toy while she's still small enough to nestle in your arms.

Unlike a standard photo album, you can easily make copies of your DVD album to give as gifts. The next time your family gathers, get out your video camera and start on a project they'll want to keep forever.

The Shoot

You're going to need a recording of everyone in your family for this video album, so a family get-together is a great place to start. That will enable you to get recordings of several people all in one day. But if you want to hold quieter interviews, and perhaps record the look of different people's homes, you can also visit each person or family separately.

tip *One side of the family or both? If you plan on making several copies of your finished album to give to your relatives as gifts, you should probably focus on one side of the family at a time. If, on the other hand, you're creating this purely for your own memories, include both sides so that you have the whole family in one place.*

The idea is to record an interview with each member of the family so that you get a sense for what they're doing with their lives at that moment. You're not just recording an event, like opening Christmas gifts or a First Communion party. You want to have a quiet moment with each person. If the word "interview" sounds too stiff, just think of it as a conversation. You want to have a quiet conversation and hear about the other person's life.

Here are a few tips for novice interviewers, to bring out natural responses from your subjects:

- Consider investing in a lavalier microphone, if you don't already have one. A lavalier mic clips to a shirt or lapel. It gives better sound than your camera's attached mic, and it's nicely out of the way so that the interview feels more casual.
- If you choose to use a handheld microphone, make sure you're the one holding it. You'll get better sound that way, and it won't distract your subject as much.
- Prepare a quiet, well-lit place to talk before your subject gets there. Move some lamps around and set up your camera first. Use three-point lighting, as described in Project 3. If your subject sees the preparation, they might get nervous about the shoot.
- Stay quiet when the subject is answering. You don't want your finished video littered with "uh-huh"s.
- Allow pauses to happen. Don't get nervous and rush the interview to fill up any dead space. When people pause and take a minute to think, they often come up with even better stories.
- Ask the person about their interests, so you get a feeling for what matters to each relative. Allow them to go into detail.
- If you're talking to an older person, ask questions about their past, about where they grew up and what careers they've had. You may think you know all about your family's history, but the answers will likely surprise you.
- When you're done with the sit-down interview, do something active. Have the interviewee give you a tour of their house or go for a ride in their car. Have a peek at an uncle's woodshop or an aunt's scrapbooking room. Get a feel for what interests them.

tip *If you're recording while you walk, hold the video camera as steady as you can. Use one arm to hold the camera and your other arm to brace your first arm against your body so the camera doesn't wobble as much.*

Editing

Once you've finished all of your recording, import the footage into your computer. Connect the USB or FireWire cable that came with the camera, and then start up your editing software. Click the option to import your video.

This project will use Adobe Premiere Elements to demonstrate, because it has handy options that make it easy to break up one video into several when you move into the DVD-creation area. With many consumer-level editing programs, you need to break your video down into separate movies and then save each one as its own file. When it comes time to create a DVD, you need to import all of those movie files into the DVD menu. But with Premiere Elements, you can keep all of your footage together during the editing process, and use markers to indicate where one section stops and another begins. It's a more natural way of working, and it lets you see all of your work at once. When you move into the DVD section, Premiere Elements treats each of these sections as different movies. You can use other markers in Premiere Elements to indicate where different scenes start and stop. This makes it far easier to create a great-looking DVD. While you're editing, you can indicate where certain stories start, for example, so that viewers can quickly jump to the area they want.

Step 1: Set Up Your Project

Once you've imported your footage into Adobe Premiere Elements, drag your file into the project area, also called the timeline area, along the bottom, as shown in Figure 8-1.

To set markers, you need to work in the timeline view. Click the Timeline button on the right side toward the bottom.

Step 2: Set Your Main Menu Marker

Premiere Elements is a powerful program, but no one said it was an easy one. Figuring out what to click to set different kinds of markers might seem hard until you realize they're all available from a right-click. Place the playhead cursor where you want to begin your first scene, say at the very beginning of the video, and then right-click anywhere on the screen. Choose the option called Set DVD Marker, as shown in Figure 8-2.

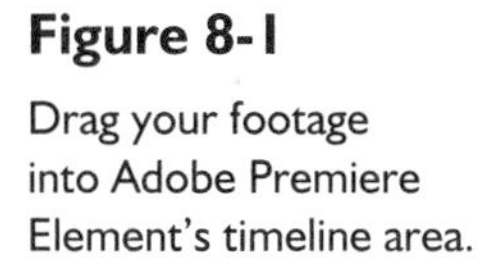

Figure 8-1

Drag your footage into Adobe Premiere Element's timeline area.

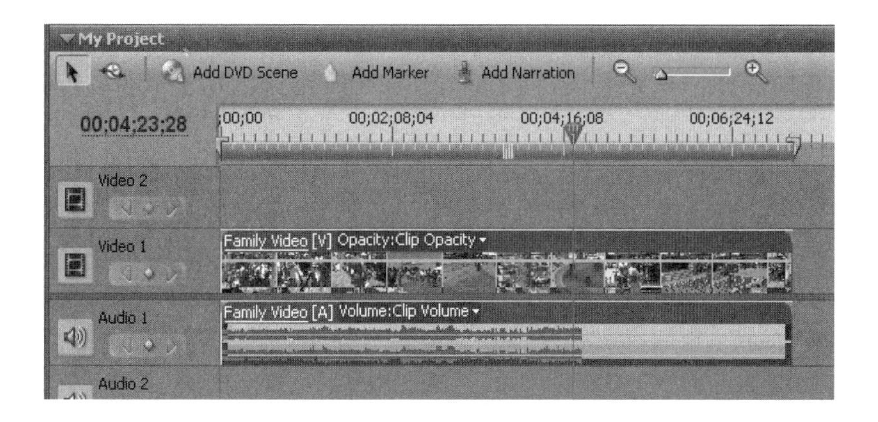

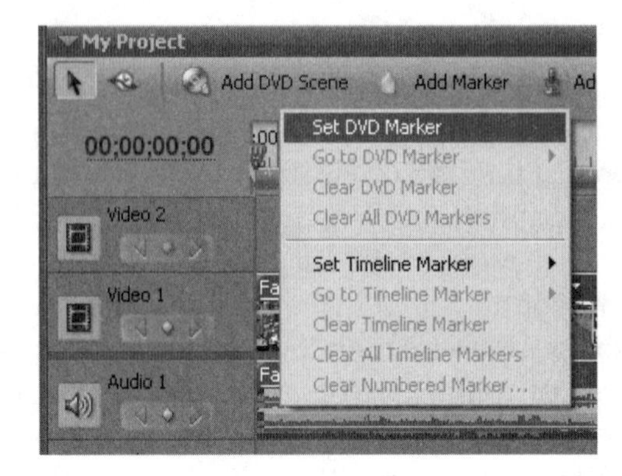

Figure 8-2
Right-click and select Set DVD Marker.

When you choose the Set DVD Marker option, the DVD Marker window opens. By default, it's set to create a scene marker, but you don't want that yet. Instead, in the Marker Type drop-down list, choose Main Menu Marker, as shown in Figure 8-3.

Premier Elements has three types of DVD markers: main markers, scene markers, and end markers. Use main markers to show how movies should be broken up for your DVD selection menu. Use scene markers to indicate different scenes within each movie, so that viewers can quickly jump to the area they want. Use end markers to tell where each movie ends. When an end marker is reached, the DVD returns to the main menu. Use end markers to mark only the ends of different movies, not the ends of individual scenes.

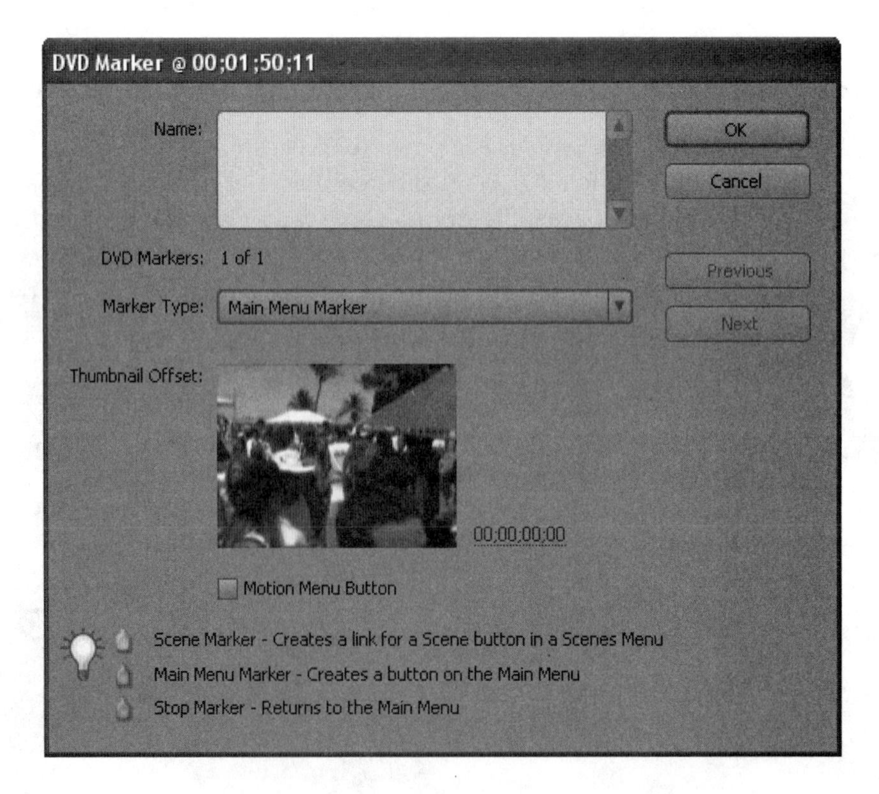

Figure 8-3
Premiere Elements' DVD Marker window

Step 3: Name Your Scene or Movie

The DVD Marker window also lets you name your movies and scenes by entering text in the Name text field at the top of the window (see Figure 8-4). Be sure to name every movie and scene. The information you enter here will be automatically copied to your DVD menu. When you're done, click OK.

Step 4: Set Your Stop Marker

You'll have interviews with several different family members all in the same string of footage, so your first step is to break the clips up by person. Place the playhead cursor at the end of the first interview—at the very last frame of that interview—and again right-click and choose Set DVD Marker.

This time, choose the option to create a stop marker from the pop-up menu. You don't need to enter any text when setting a stop marker. Click OK. You've just told Premiere Elements where your first movie ends.

Step 5: Next! ...And Repeat

Place the playhead cursor on the first frame of your second interview subject. Follow Steps 2 and 3 for all of your interview subjects, telling the program where each new movie starts.

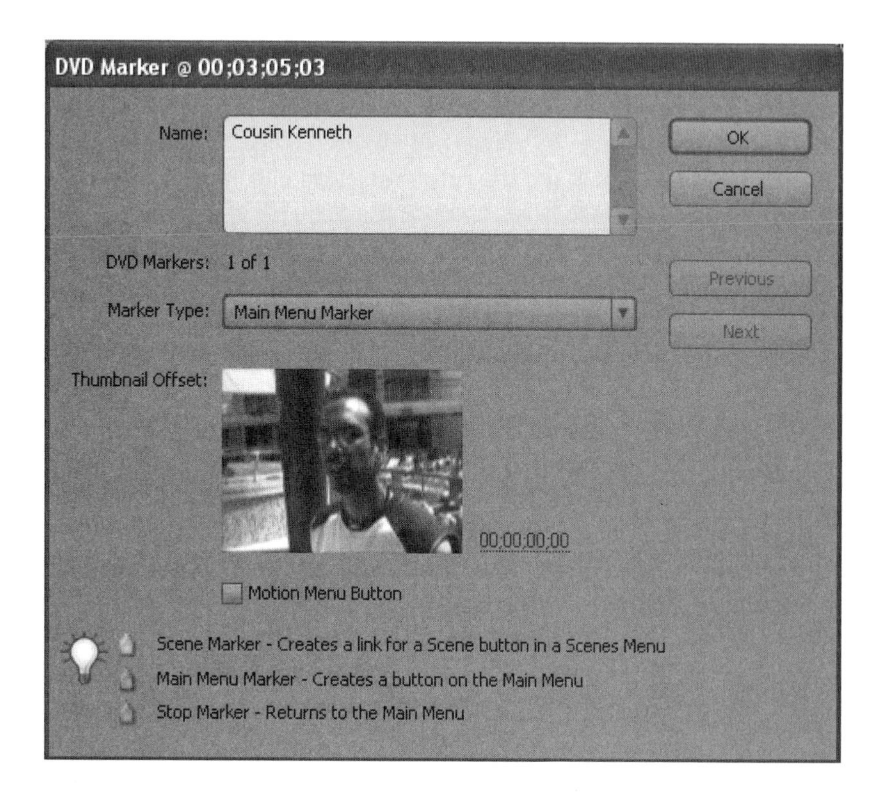

Figure 8-4
Enter text to describe each movie or scene.

Place the playhead cursor on the last frame of your second interview. Follow Step 4. Repeat this with all of the interviews in your movie, telling the program where to end each one.

tip *If you find you've entered incorrect text for a main marker or a scene marker, double-click that marker in the timeline. The DVD Marker window will open again and you can edit your text.*

Step 6: Set the Scenes

Once you've broken your footage into separate movies, you need to further break it into separate scenes. As you preview your movie in Premiere Elements' viewing area, look for areas where the subject starts new stories or a new activity. These are the places to use for scene breaks. Place your playhead cursor on the frame where a new story is about to be told. Right-click anywhere in the application window and choose Set DVD Marker. This time, stay with the default option to set a scene marker. Enter a brief text description of the scene. This description will show up in the DVD menu.

Continue with the scene marker process until you've marked all the different scenes in your movie.

Step 7: Magic Motion

The DVD Marker window always gives you the option to make motion buttons, but that can be a little confusing, since the program's DVD templates offer motion buttons only for the scene selection menus. Now that you're setting scenes, click the Motion Menu Button check box (see Figure 8-5). The timecode next to the motion option lets you set how long the button's video will play before it repeats. The video on all motion buttons will replay over and over until the viewer makes a selection. Setting the length of time is a pain, because you need to move your cursor over the timecode until you see it change to a finger over sideways arrows. Click-and-drag your cursor to the left or right to make the motion button clip longer or shorter. About ten seconds is fine. The idea is to give the viewer a taste of what they'll see if they click that button.

Step 8: Create Your DVD

When you're done setting your markers, move to the DVD-creation area of Premiere Elements by clicking the Create DVD button in the top-left corner of the main window (see Figure 8-6).

Once you're in the DVD-creation area, you'll see a list of DVD templates in the top-left corner (see Figure 8-7). Browse through them and find one you like. Consider choosing one from the Memories category. Drag the template you like into the central window. You'll see that it's automatically populated with movies and scenes based on the markers you set.

Figure 8-5

Use the motion button controls to add movement to your menu screen.

When you're satisfied with the results, put a blank DVD in your DVD drive and click the Burn DVD button in the middle of the screen.

If you want to burn several copies of your video family album, put a new blank DVD into your drive when the first burn is completed. Keep the program open and make as many as you like. If you'd like to burn more at another time, simply save your family album by choosing File | Save. You can open your saved project the next time you use Premiere Elements.

Figure 8-6

Click the Create DVD button to customize the DVD menus.

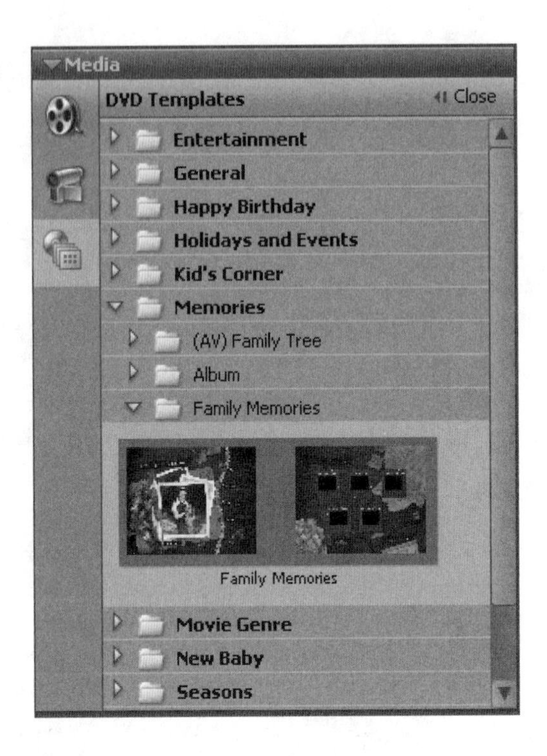

Figure 8-7

Adobe Premiere Elements comes with a wide selection of professional-looking DVD menus.

When you're done, you'll have a walking, talking family album that you can return to again and again. Families change so fast that you'll be happy you have it. Then you can "remember when" anytime you like.

Part II

Editing

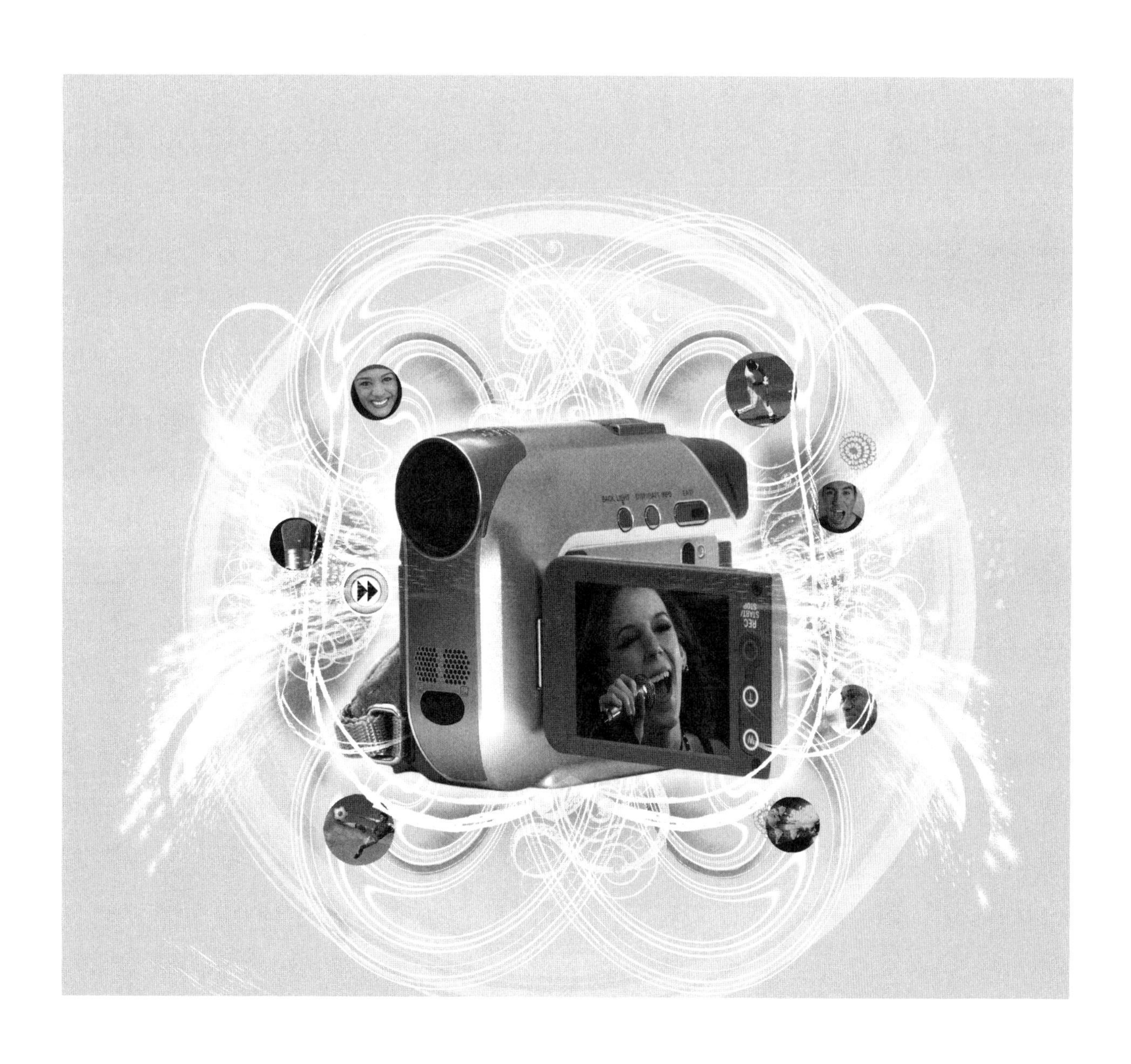

Project 9

Create a Dramatic Slideshow

What You'll Need

- **Digital video camera (any)**
- **Home computer**
- **Movie editing software**
- **Cost: $100**

Working with video software doesn't always mean working with video. Video editing programs have given new life to the slideshow—once the bane of 1950s recreation rooms and VFW halls—by letting people add motion, music, and narration. You may have never worked with slides in your life, but that isn't a problem...you don't need to have experience with slides to create an interesting slideshow.

Slideshows are a great way to provide entertainment and a fond look back during events such as wedding receptions and anniversary dinners, letting you dust off your old family photos and share them. They're also great for sharing your latest digital camera pics. How are you showing your digital photographs to friends? Using the viewing screen on the back of the camera? E-mailing the occasional good shot? You're probably not printing copies of them; using a photo printer or uploading your photos to an online printing lab is supposed to be easy, but it usually seems like a nuisance. It's far simpler and more fun to create a slideshow and burn it to a DVD. This project, will take you through a few ways to create slideshows and show you how to share your creations when you're done.

Scanning Old Photos

Scanning old photos into your computer for a digital slideshow is a fun way to breathe new life into your images. Here's how to get them looking their best.

Step 1: Clean the Screen

Clean off your scanner glass before you start, and check the photos themselves for dirt or dust. A little cleaning at the start will give you better scans (see Figure 9-1).

Figure 9-1

Clean the scanner glass before you start.

Step 2: Adjust the Resolution

If you're scanning images in just for a slideshow, scanning them at 300 dpi (dots per inch) is a good all-purpose resolution. If you're going to print them, especially in larger sizes, use a higher resolution, such as 600 dpi.

Step 3: Remove the Yellow

If you plan on scanning a lot of old photos, invest in photo editing software that can make them look brand new again. Adobe Photoshop can easily remove the yellow cast from an old photo; select Image | Adjustments | Auto Levels.

tip *If you're shooting digital photos expressly to add to a slideshow, set the size ratio to match the ratio of the screen you'll be using. So, for example, if you're going to show your slideshow on a widescreen TV or monitor, shoot pictures with the 16:9 resolution. If you'll be using a standard set or monitor, shoot at 4:3.*

Creating Slideshows with Apple iMovie and iPhoto

Apple includes the iLife suite of media applications free with every Macintosh computer, and the suite includes two different ways to create slideshows. One is easier, but the other offers more control.

Apple iPhoto

Using iPhoto to create a slideshow is a snap. The program is Apple's photo storage application, so if you're taking pictures with a digital camera, you're likely already using it to import and organize your shots. It doesn't offer as many options as iMovie does, but it makes slideshow creation a breeze. Here's how to do it.

Step 1: Organize Your Photos

Your first step is to organize in one place all the photos you want to use, so that you can easily grab them all. Create a new folder by clicking the "+" button in the lower-left corner of iPhoto. From the pop-up window, select New Album, give the folder a name, and click Create. Return to your main library view by clicking Library in the top left. Now, drag all the pictures you want to use into the new folder, which you'll see on the left. When that's done, click the new folder's name on the left to see the contents. Drag the photos into the order in which you want to present them.

Step 2: Configure Slideshow Settings

To instantly create a slideshow from your pics, click the Slideshow button along the bottom of the screen. You'll now see new options along the bottom (see Figure 9-2) that let you customize your slideshow. You can add color effects, slide transitions, or the Ken Burns motion effect.

Figure 9-2
Apple iPhoto's slideshow options

note *All the programs we're looking at let you add motion to your slideshows, but they call the option by different names. Apple calls it the Ken Burns Effect, referring to the esteemed documentarian known for creating drama from historic photos, while most others call it pan-and-zoom. It refers to the same effect in both cases.*

iPhoto automatically uses the Ken Burns effect even if you don't have it selected. To turn it off, click Settings and uncheck Automatic Ken Burns Effect. Customizing the Ken Burns effect isn't as clear as it should be. To do so, check the box for the effect in the bottom menu and make sure the slider control is set to Start. This is how your image will first look. To change the size, drag the resizing control on the lower right, and then reposition the image by grabbing it with your cursor and moving it. Next, set the slider to End; this is how your image will end up looking. Again, drag the resizing control to adjust what the viewers will see, if you want to show less than the full image. To make sure you've got what you want, click Preview.

tip *Want to quickly add an effect to all of your slideshow photos with iPhoto? Highlight the pictures you want to use in the layout at the top of the screen, and then select the effect from the pop-up window along the bottom. If the pictures you want to highlight aren't next to each other, press the COMMAND key first (the one with the cloverleaf pattern) and then click the right photos.*

Step 3: Add Beautiful Sounds

iPhoto lets you grab music tracks from your iTunes library. Click the Music button to grab a song or two. iPhoto gives you the handy option of making sure your slideshow exactly matches the length of your soundtrack. To set this, click Settings along the bottom and choose Fit Slideshow to Music (see Figure 9-3).

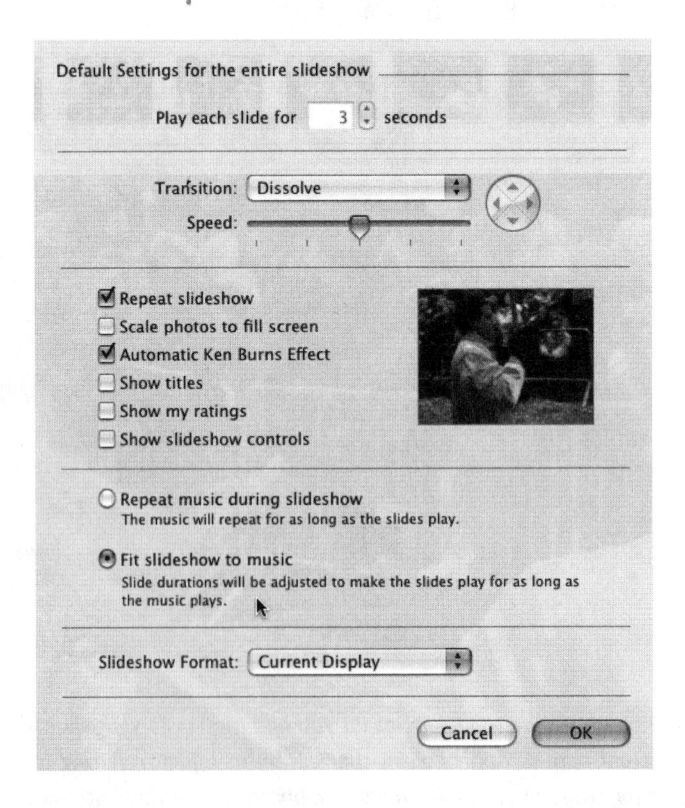

Figure 9-3
Apple iPhoto lets you add a soundtrack and set your slideshow to match the length of the songs.

tip *You can still edit your photos while creating a slideshow in iPhoto. Double-click your photo to enter the Edit menu. When finished, click Done to return to the slideshow options.*

Step 4: Export Your Creation

When you're finished, you can save a copy of the slideshow to your desktop by choosing File | Export. You'll get a choice of three sizes. Saving to your desktop produces an MOV file that can be viewed by anyone using QuickTime. You can also send your slideshow to iDVD by selecting that option from the Share pull-down menu. In iDVD, you can then burn your slideshow to a DVD. Note that selecting Email from the Share menu sends only the picture you currently have highlighted, not the entire slideshow. If you want to e-mail a slideshow, export it as an MOV file first, and then attach that file to an e-mail message.

Apple iMovie

Working in iMovie is a bit more involved, but you get more control over your final results.

Step 1: Select Pics

Open iMovie, click the Media button at the bottom, and then click the Photos button at the top. You'll see all the photos you have stored in iPhoto. Select the photos you want and drag them into the timeline (see Figure 9-4).

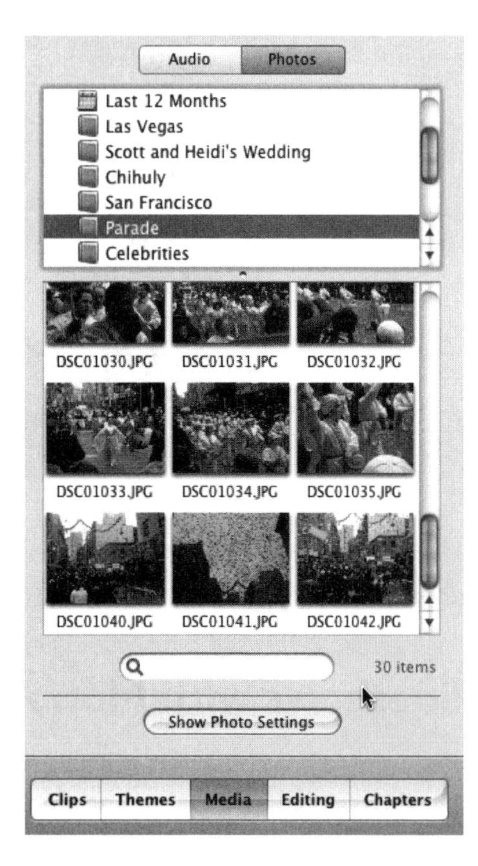

Figure 9-4
Choose the photos you want to use from iPhoto's image browser.

Step 2: Too Much Motion?

iMovie adds the Ken Burns effect automatically to imported photos. To turn this off, select a photo from the photo browser before import and click Show Photo Settings. Uncheck the Ken Burns Effect check box.

Step 3: Make Your Photos Fit the Screen

If your photos are at a high resolution, they'll appear too large in iMovie. To make sure they fit the screen, click Show Photo Settings and use the resizing bar to adjust the fit. If you don't want motion on a photo, make sure the Start and End images are the same.

Step 4: Add Finishing Touches

iMovie sees your slideshow as scenes in a movie rather than as static images. You can drag in transitions, music tracks, and titles just as you normally would.

Step 5: Export the Slideshow

Because you're using iMovie, you have stronger export options than you would if you were using iPhoto. You can save a QuickTime file to your desktop, e-mail an MOV file, send your file to iDVD to burn it to a disc, format your file for the iPod, or load the movie onto your video camera. You'll find all of these options under the Share menu, as shown in Figure 9-5.

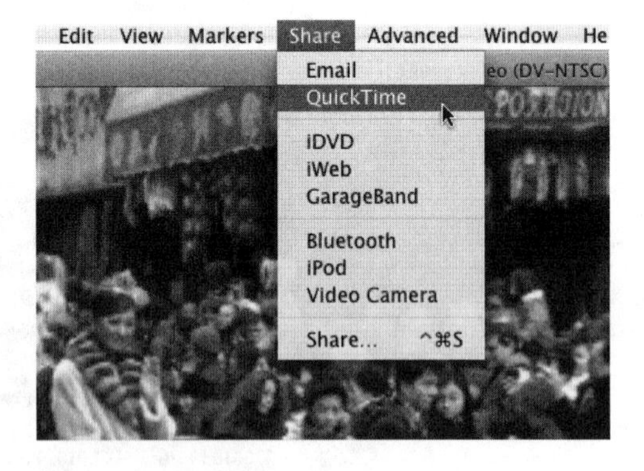

Figure 9-5

Apple iMovie gives you a long list of export options.

tip *Want to add an image from a video to a slideshow? You can create a still image of any frame by placing the playhead on the exact frame you want and choosing Edit | Create Still Frame.*

Creating Slideshows with Motion Pictures

If you're using a Mac and you want to create really stunning slideshows, invest in an application called Motion Pictures, from Roxio (www.roxio.com). Motion Pictures currently sells for $39 on its own, but it also comes with Roxio's excellent disc-burning suite, Toast, which is $79 after rebate. It lets you do a few tricks that iMovie and iPhoto haven't learned yet:

- Open Motion Pictures and you'll see your iPhoto pics listed in the left column with the slideshow area in the middle. Drag the pictures you want into the slideshow area. Select one of the slides in your slideshow and click the Group Pictures button at the top of the window. You'll see that Motion Pictures lets you display up to six images on each screen, for a much more interesting slideshow (see Figure 9-6).

Figure 9-6 Use Motion Pictures to create dramatic slideshows on the Mac.

- Select one of the images in your slideshow and click the Motion button at the top of the window. You'll see a well-organized workspace that shows how your image will start and end up. You can create pan-and-zoom effects, as with other programs, and also add rotation, so that your images rotate a bit on the screen.

Creating Slideshows with Adobe Premiere Elements and Photoshop Elements

You could easily create a slideshow with Premiere Elements, but a better idea is to make one in Photoshop Elements and import it. The two products are sold bundled together, so it's a good value if you need a strong photo editor and a strong video editor. For slideshow creation, Photoshop Elements has better tools, and you can easily import your creation into Premiere Elements to add it to a longer movie or DVD project.

Step 1: Import Photos

If you've never used Photoshop Elements, you need to add some photos first. Open the program and choose File | Get Photos.

Step 2: Set Your Preferences

Choose Create | Slideshow to open the Preferences panel, which lets you set the basic slideshow options, such as how long each photo should display, what kind of transition or transitions you'd like to use, the background color, whether or not you want to use pan-and-zoom effects, and more. Click OK when you're done.

Step 3: Use the Slide Show Editor

You'll next see the program's Slide Show Editor, shown in Figure 9-7. Add some images by clicking the green Add Media button. Choose Photos and Videos from Organizer to see pictures you've already added to Photoshop Elements. Put a check mark by the ones you want to use and click Done.

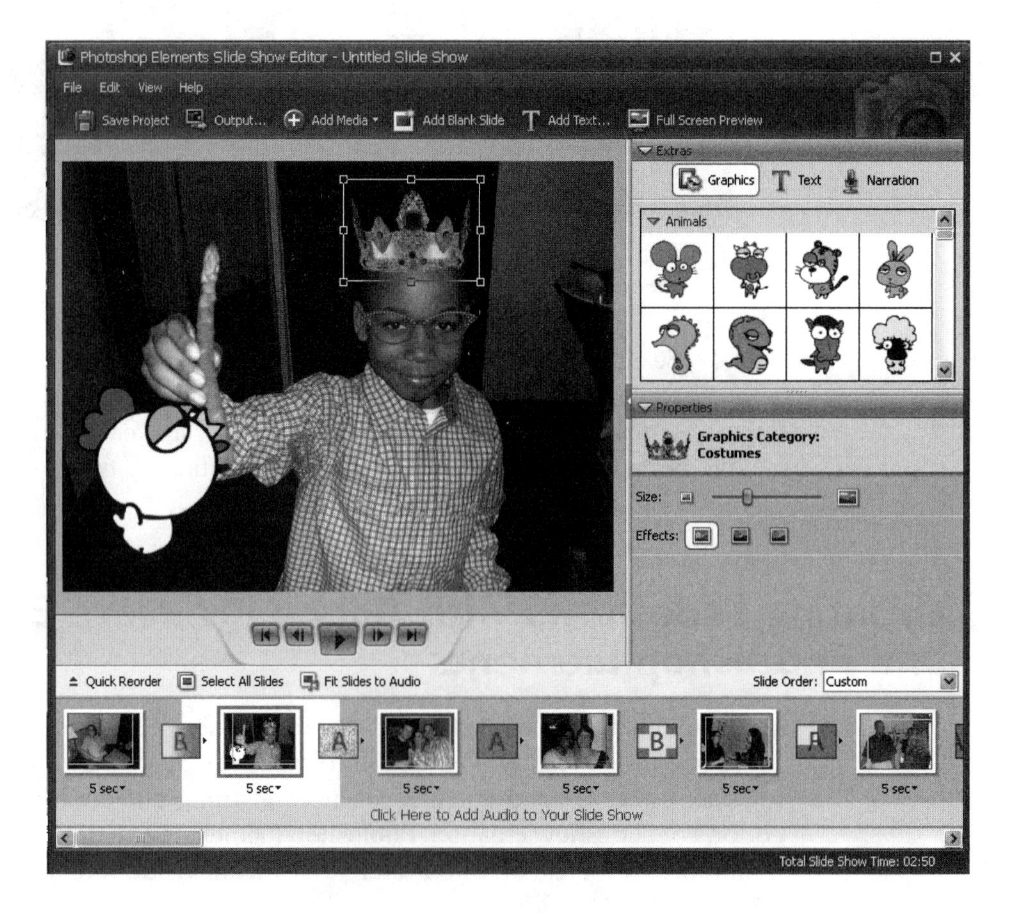

Figure 9-7 Adobe Photoshop Elements has some especially humorous slideshow options.

Step 4: Add Creative Touches

Photoshop Elements' available options make it the easiest and most enjoyable slideshow creator around. You can add graphics from the menu on the right, such as cartoon animals and speech balloons, by dragging them onto your slides. Drag in a speech balloon and then click the Text button at the top right to add words to it. Type the words you want and then drag them on top of the speech balloon. It's a great way to get your guests laughing.

Step 5: Make it Move

Photoshop Elements makes it easy to see what you're doing with pan-and-zoom effects, by putting a green outline around the beginning state of your photos and a red outline around the end state. To adjust the motion, select the photo you want from the list along the bottom and then click either the start or end image from the right side. In the large workspace window, resize the green or red outlines to where you want them to be.

tip *Your slideshows don't have to be purely static images. Throw in video clips here and there to add interest.*

Step 6: Add Narration

Adding narration is also simple with Photoshop Elements. Click the image you want from the list along the bottom and select the narration option from the top right. Click the red Record button and speak into your computer's microphone.

Step 7: Export the Slideshow

When you've got your slideshow just how you want it, click the Output button near the top-left corner. From the options window, you can choose to export your slideshow to Premiere Elements, where you can combine it with video, or save it, burn it to a disc, or e-mail it. You can even make it into a PDF file, in case you want to make a printed copy.

When you're using a pan-and-zoom effect, do it with a purpose. Zoom in to show detail or zoom out to reveal a scene. You can also pan across a shot, to give a feeling of space. But never include an effect just for the sake of adding movement—and never accept automatic pan-and-zoom settings.

Slideshows aren't dull anymore. With a little practice, you'll be making ones that are just as interesting and full of movement as an actual video. Try out one of the programs discussed in this project and give your photos a whole new life.

Project 10

Add Terrific Titles and Creative Credits

What You'll Need

- **Digital video camera (any)**
- **Home computer**
- **Movie editing software**
- **Cost: $100**

If you're only slapping a basic title on your movies and not doing anything at all for end credits, you're missing out on an opportunity to be creative and delight your viewers. Spend some time thinking about the credits for your movie, because there's no reason they have to be dull. With a little planning, you can come up with something that wows your audience and sets the mood for your film. You might want to use objects around your house as props while you film the credits, or you could use the built-in title and credit controls in your editing program. Today's programs make basic credit creation easy, and even include effects that mimic the credits from well-known movies. I'll teach you a few advanced techniques so that you can do something really special with your credits, leaving your audience wondering how you pulled it off.

Using Household Objects for Credits

Did you see the movie *Napoleon Dynamite*? If you did, you certainly remember the eye-catching opening credits, which were spelled out on foods—foods that were later eaten in scenes in the movie.

That's a perfect example of how you can create memorable credits using objects you already have around your house. Consider these possible openings:

- **New baby video** Spell out the title and opening credits using wooden blocks. Pose stuffed animals and other toys around the blocks. Film it in a nursery, with items such as a crib or children's wallpaper out of focus behind the blocks.

- **Travelogue** Collect postcards of the places you visit and then paste them into an album. Type out titles and credits on your computer and print them out. Cut out the titles and credits and paste them around your postcards. Next, film the album while you open it and turn the pages. Each page gives a postcard view and a little bit more of the credits.
- **Romantic movie** Film a bouquet of roses in a vase, in soft lighting. Zoom in to show a close-up of the card. Instead of the usual "to" and "from," show the name of the movie.
- **Horror movie** Write the credits on a mirror using lipstick or something that looks like blood, such as ketchup with breadcrumbs used as a thickener. Be careful when you film the mirror that you don't end up in the shots.
- **Wedding video** Use a word processing program on your computer to create a fake wedding announcement, full of flowery script. Print it out and scatter rose petals around it.

Adding Titles and Credits with Apple iMovie

Apple iMovie makes it easy to create some surprisingly sophisticated titles.

Step 1: Learn Your Way Around

To find the title and credit options, click the Editing button at the bottom of the main iMovie window and click Titles at the top.

The interface can be a bit confusing at first, but it's simple once you know your way around. Title styles are listed at the top, as shown in Figure 10-1. If there's an inverted triangle next to a title name, you can click it to see variations on that style. Below the title styles are boxes in which you can enter the onscreen text, and a color-choosing box where you pick the color for your titles or credits. Below that you can choose a font and the lettering size, and below that you can choose the length of time that your credits will display.

Step 2: Begin with Black

By default, iMovie credits display over the opening seconds of your movie. If you prefer that the credits stand on their own, click the Over Black check box toward the bottom of the window.

Step 3: Make Additions and Changes

When you're ready to add titles or credits, you have two options: you can click the Add button, or you can drag the credit in by clicking the "T" box to the left of the credit style's name and dragging it into the timeline.

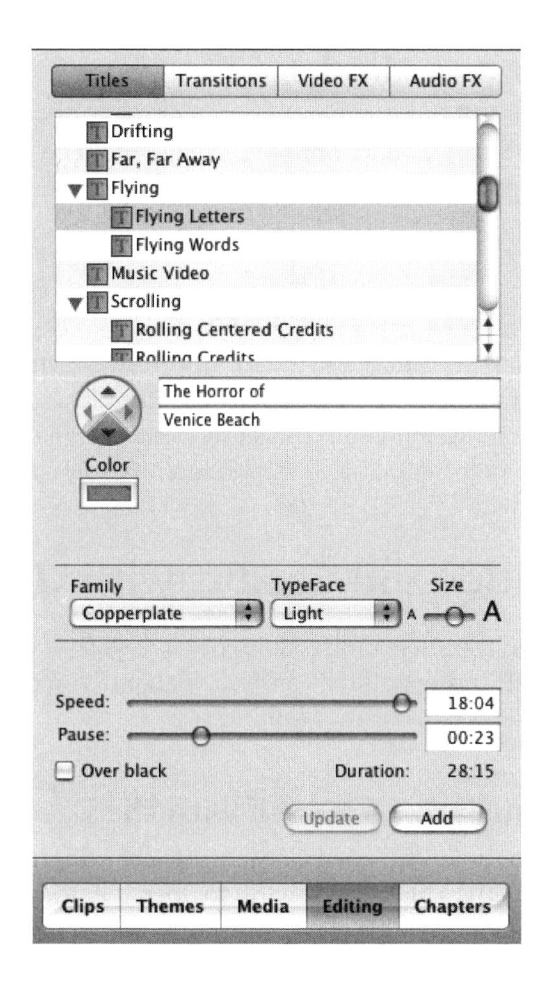

Figure 10-1

Apple iMovie's titles and credits let you create a professional look with little effort.

Be careful when making edits to your titles or credits. Always click the Update button in the lower right when you're done (see Figure 10-2), or else you'll lose your work.

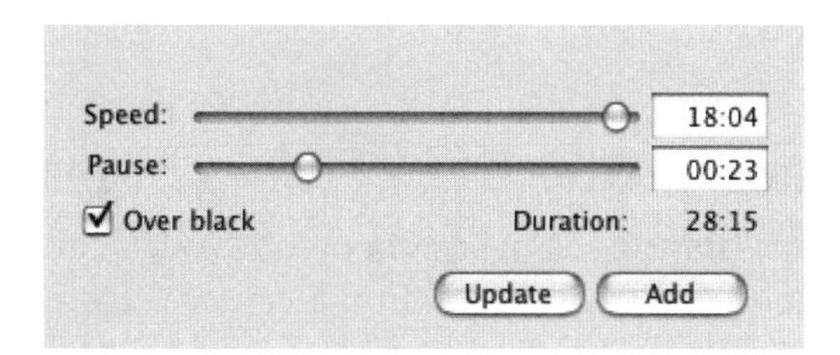

Figure 10-2

Be sure to click Update to save your changes.

Step 4: Choose Title Style

Play around with the various title styles before you make your selection. There are two Typing styles, one that mimics the look of old typewriters and one that mimics the look of a computer screen. You could easily use these within a movie as well, such as when a character is typing a letter.

Step 5: Create a Mood

Think about the mood you want your viewers to be in. The Bouncing styles, where the letters bounce onto the screen, might be good for a video of a family outing, while Wipe, which shows the title wiped and then unwiped on the screen, carries a feeling of menace.

Step 6: Consider Special Uses

Some titles in the list aren't meant for beginning or ending a movie, but rather for special uses inside of one. Check out Music Video, which lets you create a block of text in the lower-left corner of the screen, just like an MTV music video. Stripe Subtitle is a polished-looking way to identify someone during an onscreen interview.

Adding Titles and Credits with Ulead VideoStudio

Don't let first impressions deceive you. VideoStudio's title options look skimpy at first, and not all that professional, but you can do impressive things with them once you know your way around.

Step 1: Familiarize Yourself with the Layout

After you've loaded a movie into the editing timeline along the bottom of the screen, click the Title tab at the top to open the title options. VideoStudio doesn't give you as many preset title effects as iMovie and Premiere Elements give you, but it gives you a lot of control over the outcome. The various title styles are on the right side, while the title editing controls are on the left.

When you first open the title controls, you'll see a message on the middle left telling you how to enter text. It seems as if this message is just filling up unneeded space, but it's actually hiding some of the editing controls. Click the double-arrow symbol beside the message to see the full list of title editing options, shown in Figure 10-3.

Step 2: Select a Title

Your first step is to select one of the basic title styles from the right. When you click one, you'll see it in full in the center preview window. Click the text in the preview window to put an editing box around the text, and enter your own title.

Step 3: Juice Up Your Title

After you've entered your title, you're ready to play with the real power of VideoStudio's title controls. Click the Animation tab. Some of the titles have two different editing areas. When this is true, you can control the animation for each part separately. Select one of the text boxes and then click Apply Animation in the left column. The drop-down menu

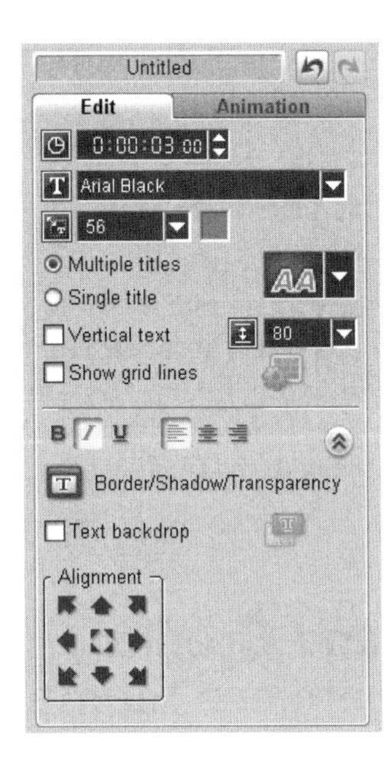

Figure 10-3

Click the double-arrow symbol to see the full title editing options.

next to the check box lets you select different types of animation (such as Fly or Swing); select one and you'll see varieties of that movement below. Whichever style you click is automatically associated with the text box you have highlighted.

Step 4: Edit Your Title

Click the Edit tab to see the text editing options. VideoStudio has a handy visual font menu, so that you can see what each font looks like before you select it. Click the arrow next to the font box to see the styles available on your computer (see Figure 10-4).

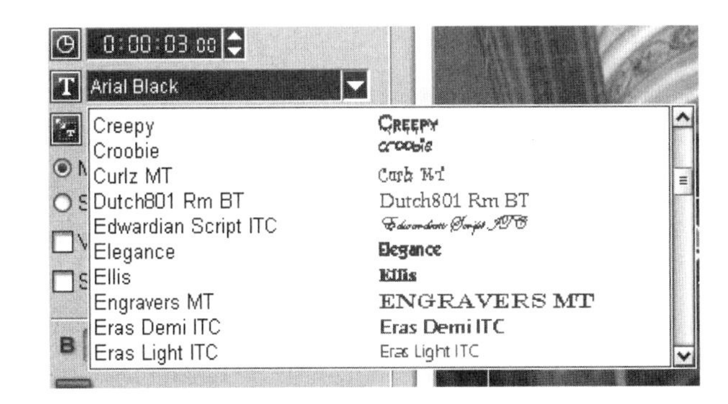

Figure 10-4

VideoStudio's visual font menu shows you what each font looks like.

Step 5: Make Additions

Once you have a title that you like, figuring out how to load it into your movie can be confusing. There's no Add button that you click to move it to the playhead. Instead, you click the title style that you first chose from the right column, and drag it into the title area of the timeline. This is the area marked with a "T."

You can simply drag your title clip in the timeline to make it display for a shorter or longer period of time.

Step 6: Check Out the Library

Once you've fine-tuned a title style in VideoStudio, save it in case you want to use it again. Use the drop-down menu in the upper right to enter your personal library area (you'll have to make a library if you don't already have one). Then, drag your title from the timeline into the library. You can now easily use your customized style anytime.

Sadly, VideoStudio doesn't have end credit options. You can make great-looking opening titles, but no closing credits.

Adding Titles and Credits with Adobe Premiere Elements

Premiere Elements lets you make far more interesting and professional titles and credits than any other consumer-level program, but the process is also more complex than in other consumer-level programs. If you find yourself stuck—and you just might—don't give up. Search through the included help files and you'll likely find a solution.

Step 1: Open the Titles Menu

With a video clip open in your timeline, click the Titles menu a quarter of the way down on the left side. You'll see many different title options, broken up into categories (see Figure 10-5). Click the triangle next to any of the categories to see the contents. Don't be discouraged if you don't see anything right away; the categories might take a minute to load.

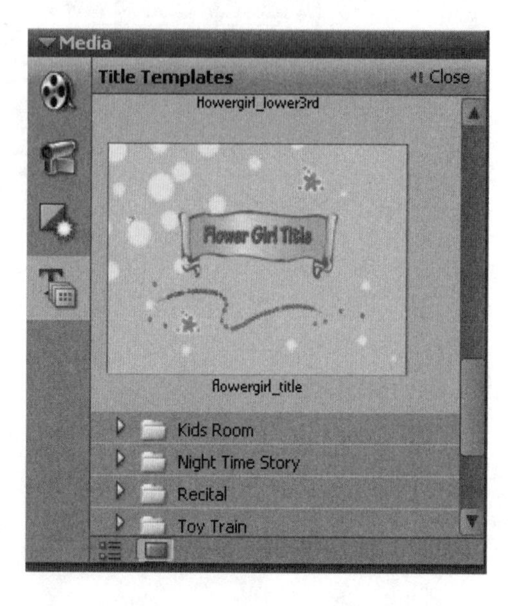

Figure 10-5

Adobe Premiere Elements has the best-looking title options of any consumer-level editor.

Step 2: Add a Title

Adding a title can be a little tricky. If you want to show the title before your video, you can simply drag it into the video track of your timeline. But if you want it to play over your video, you need to drag it into a separate track of the timeline. If you put it in the same track and try to merge it with the existing video, you'll only cover the title up and it won't be visible.

Step 3: Change the Text

Double-click the title in the timeline to call it up in the workspace. You edit the text directly in the preview window, so click the onscreen text.

tip *You'll see two boxes around your title, showing you the safe area that will display on different television screens. The amount of your movie that will display can vary from set to set, but you'll be safe if you keep your title within the inner box. If your title gets too wide for the space, you can force it to wrap to a new line by right-clicking the text and checking Word Wrap.*

Step 4: Insert an Image

One of the novel things you can do with Premiere Elements' titles is add images. You'll see a small white box with a red triangle midway down on the right side. This is the Add Image button. Click it and select an image from your hard drive.

tip *Images from digital cameras initially display far too large on your screen, and cover the whole area. Drag the image until you can reach the resizing boxes on the edges, and then drag them to make the image smaller. To keep the proper proportions while you resize, hold down the shift key as you do it.*

Step 5: Fade to Black

You can add an impressive fade-in or fade-out effect to your titles or credits by clicking the clip once in your timeline to open the Properties menu, shown in Figure 10-6 (you may need to click something else first, if you're already editing the titles). Click either the Fade In or Fade Out button to add one of those effects.

Creating a Freeze Frame

You can create a freeze frame effect with any editing program. It's a fun and unexpected way to end your movie.

Step 1: Pick Your Spot

Decide what you want the last image of your movie to be, and place the playhead on that exact frame. Click the button that lets you split a video into two segments.

Figure 10-6

Make your credits fade in or out with Premiere Elements.

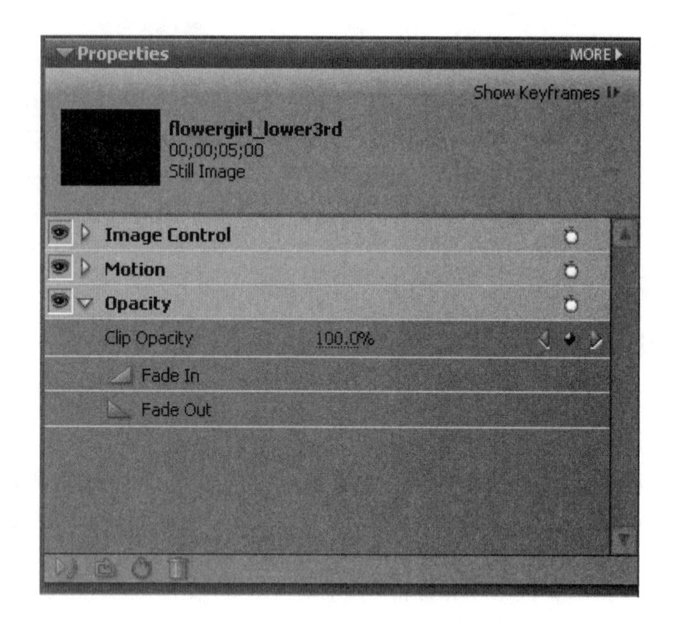

Step 2: Take Out the Trash

Click the video segment after your split to select it, and then click the Delete button. Your video now ends with the frame you selected.

Step 3: Freeze!

Put your playhead on that last frame, if it isn't already there, and click the button to make a still image from that frame. In Premiere Elements, for example, you'll see a camera-shaped button in the upper right that's actually called Freeze Frame (see Figure 10-7).

Figure 10-7

Click the Freeze Frame button to add a freeze frame ending to your work.

Step 4: Stretch That Shot

Insert the still image at the end of your movie. Drag its right edge in the timeline to stretch it out and make it longer.

Step 5: Let the Credits Roll

Finally, go to your program's title and credit options and add text over that still image. To your viewers, it will look like your movie simply freezes on that last frame while the credits roll.

Framing your movies with perfect titles and credits will make them look that much more impressive. Follow these instructions and you'll start your movies on a strong note and end them with a professional flourish.

Project 11

Giants Attack: Create a Monster Movie

What You'll Need

- **Digital video camera (any)**
- **Home computer**
- **Movie editing software**
- **Green screen supplies**
- **Cost: $100 or more**

It's amazing what consumer-level editing software can do nowadays. While the early versions of programs focused on editing, adding transitions, and plugging in audio, today's programs offer surprisingly sophisticated effects. The best example is chroma key editing, which is another way of saying blue-screen or green-screen editing. Chroma key lets you remove a color from your video, so that a still image or a second video can show through that area. It's an effect that you could find only in professional-level products until fairly recently, and it's still not available in every editing program. If having it is important to you, check the software carefully before you buy.

We're going to use chroma key effects to make a homemade monster movie in which the family dog (or other pet) grows to a giant size and rampages through the neighborhood. Of course, after you learn how to use chroma key, you can use it for anything, such as to show yourself in Paris, on the moon, or anywhere else. You can even use it to give a weather report—just like the weather people on television do—if you're working on the family newscast project from Project 2.

Creating a Filming Area

Before you can shoot your masterpiece, you need a green-screen area for recording. (A green screen is used in this project, but blue also works well.) If you're working on a newscast video, you can get away with creating a small green screen above a prop desk, but for full-body shooting, you're going to need more space.

If you've ever seen professionals making a newscast or a special-effects movie, you know that they use bright green or blue backdrops. The reason is partly that those bright colors show up distinctly on the video, so they're easier for the computer to separate and remove than dim colors. So why don't pros use bright red or orange? Because those colors are more likely to show up in skin tones. For the effect to work well, you need the background color to be completely unlike the other colors in the shot.

When you are preparing to construct your filming area, you have to decide how professional you want your finished movie to look, and how much money you're willing to spend. If you're going to enter your movie in independent film festivals, you'll want to visit a professional photo and video store to pick up a frame and a chroma key muslin. They actually don't cost that much; you can get a good system for between $100 and $300, depending on the size. Another option, if you're serious about movie-making, is chroma key paint. You can create a filming area in the basement by painting a section with special green or blue paint. It currently costs about $60 per gallon, and is also available through professional photography and video stores. If you decide to do this, paint a section of the surrounding walls, ceiling, and floor, so that your actors are free to move around.

For our purposes, though, that's all overkill. You can create a simple shooting area for far less money. One option is to pick up several brightly colored poster boards from an office supply store (all the same color, of course). Line a wall with them and make the edges meet at tightly as possible. You don't want visible cracks between them, because that could spoil the effect during editing. Don't overlap the edges, because that will likely cast a shadow, which will be even more visible. Put poster boards down on the floor next to the wall as well.

You could also hang up a sheet to use as a backdrop, but you need to be careful about how you do it. First, the sheet should have a bright and bold color; as mentioned earlier, blue and green work better because they're distinct from skin tones. The sheet also needs to have strong color saturation. That means it can't have white tones in it. It needs to be a bold, bright, solid color. (It'll be hard to find in stores, because that's one ugly sheet.)

tip *If you're using a sheet, be sure to pick a fabric that isn't reflective. Shiny spots will cause problems during editing. That's why professional backdrops are traditionally made from muslin.*

Another challenge when working with a sheet is that you need it pulled taut from all directions, so that there are no wrinkles or areas that could cast shadows. You could pin it to a hanging drapery or string it tightly between two ladders. Do whatever works for your workspace, but remember that it has to be perfectly flat, with no bulges or wrinkles.

Lighting is also a special concern when doing chroma key effects. Usually when you're filming, you need to focus your attention on lighting your actors. When you're shooting chroma key, you need to pay just as much attention to how you light your backdrop. That's because the entire surface needs to be one bright, even color. The more even it is, the easier it will be for your software to isolate that color later on.

Place your most powerful light on one side of the backdrop, lighting it from an angle. Put a second, softer light on the other side, to make sure everything is covered and to bathe the surface in a less-harsh glow.

tip *Don't let your actors wear clothing that is close in color to the background color, or else their clothing will "disappear" during editing.*

You will use separate lighting for the actors, so you need several lights to do chroma key effects well. You want to be sure the actors don't cast a shadow on the backdrop, so place them well in front of it, about five to ten feet. For this project, though, we're recording a family pet. The pet won't cast as long a shadow, so you can put him or her closer to the backdrop.

Step 1: Place the Camera

Put your camera down low, perhaps right on the floor. Think about the final movie and the view you want your audience to have. You want the animal to appear enormous, so shoot from below to make it seem even larger. You don't want it to completely fill the screen, though. Leave room for the backdrop and the other actors you're going to add in.

Step 2: Action!

Have your pet move across the frame while you film. You'll use this shot later, to show the giant pet thumping down the street.

Completing the Rest of the Shoot

For the rest of your shoot, you want scenes that show how the family pet grew to a giant size (maybe it was chemicals from sister's hair gel; maybe the pet broke open a bottle of Miracle-Gro), and scenes that show the family trying to get the pet back to its normal size. This project won't go into detail, because these are creative decisions that you can make on your own. Only one chroma key shot is presented here, but you're free to add others. Have fun creating scenes that show the destruction a giant pet could cause.

You need an exterior shot to use with the chroma key shot you just recorded. On a sunny day, set up your camera a good distance from your house, probably across the street. Frame the house off to one side. Think about how you recorded the pet and how the blended scenes will look. When you yell "Action!" have your actors run across the scene, looking back (and up) in horror.

tip *For added realism, the light source direction for your exterior shot should match the direction of the light source for the chroma key shot.*

Creating a Chroma Key Effect with Adobe Premiere Elements

This project will demonstrate how to create a chroma key effect with only two programs, since Apple iMovie, as of this writing, doesn't offer that capability. Perhaps a new version will offer it by the time you read this.

Premiere Elements has the most sophisticated chroma key controls of the consumer-level editors, although if you rely on the included manual, you might not even know they're there. That's because the included manual doesn't show how to create chroma key effects; but oddly enough the large how-to book that Adobe Press sells does. Luckily for you, so does the book you're holding—and this one is cheaper.

Step 1: Import Your Footage

Start by importing your footage into Premiere Elements. You'll need both segments you shot for the chroma key: the interior and exterior shots.

Step 2: Timeline Configuration

Edit the two segments to about the same length and place them into the program's timeline, along the bottom. Put the chroma key shot in the track labeled Video 2 and the exterior shot in the track labeled Video 1 (see Figure 11-1).

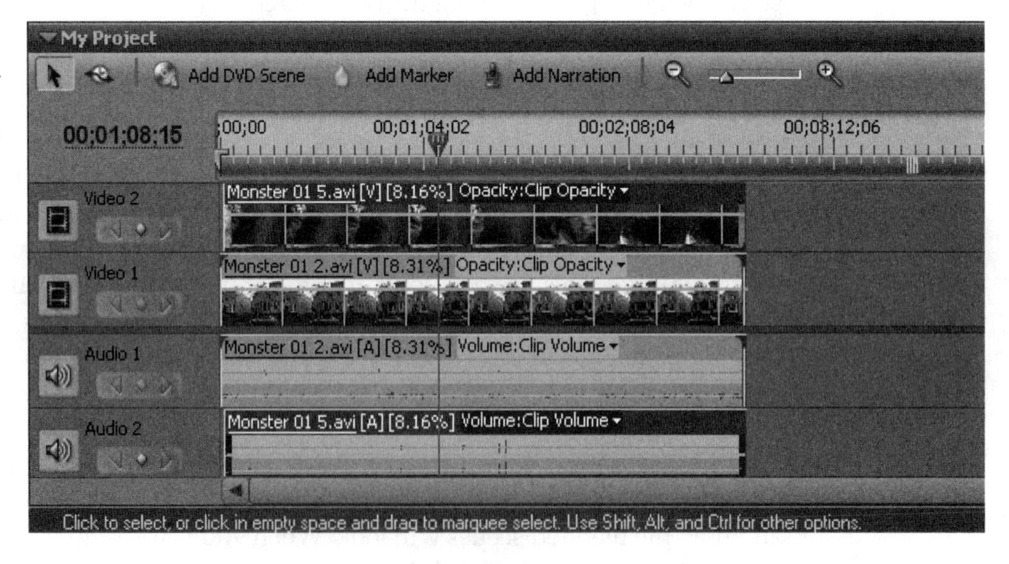

Figure 11-1

Place your two film segments in two different tracks, with the chroma key shot in the track labeled Video 2.

note *You can easily use a still image during chroma key editing—for example, to show yourself speaking in what appears to be an exotic location. In this case, import the still shot into your program and place it in the Video 1 track. Drag the right edge of the picture so that it's as long as the video in the track above it.*

Step 3: Select the Effect

Click the chroma key track. Open the Effects and Transitions menu on the left side by clicking the icon that looks like a purple box with a sparkle in front of it. Click the triangle next to Video Effects and then click the triangle next to the word Keying. This shows you all the key frame effects. Scroll down until you see Chroma Key (see Figure 11-2). Click-and-drag the Chroma Key icon down to the video in Video 2 and release it.

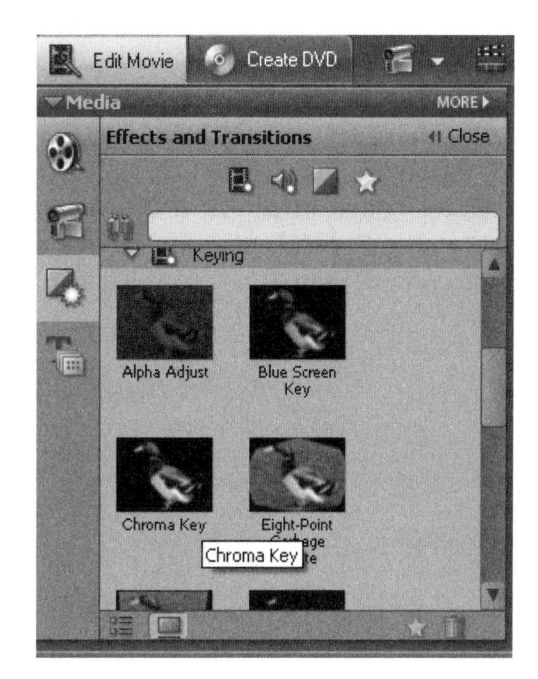

Figure 11-2
Select the Chroma Key effect from the Effects and Transitions menu.

Step 4: Choose Your Settings

Click the video in Video 2 once and you should see an icon representing that video in the top right of the screen, in the Properties menu. You should see a bar labeled Chroma Key in this menu. Scroll down if you don't see it.

Step 5: Use the Color Picker

Your first step in adjusting the settings is to use the color picker. Click the eyedropper tool (see Figure 11-3), and use it to select the backdrop color from your video in the center preview window. As soon as you do so, you should see that color disappear from the video and the footage from the track in Video 1 show through.

Step 6: Customize It

If the results don't look perfect right off the bat, you'll need to adjust the Chroma Key properties a little bit (see Figure 11-4). Start with Similarity. Click and hold on

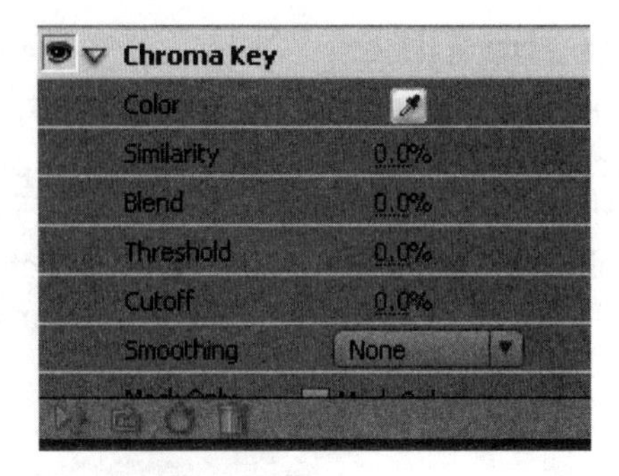

Figure 11-3
Use the eyedropper to select your chroma key background color.

the 0.0% measure for Similarity, and then drag your cursor to the right or left to raise or lower the amount. Increasing Similarity means that Premiere Elements will also key out colors that are similar to the color you picked in Step 5, even if they are not a perfect match. Experiment with the other controls to find a level that looks convincing with your pet's fur.

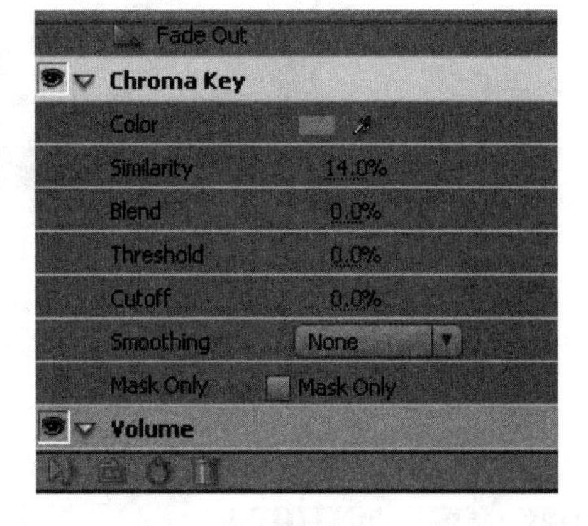

Figure 11-4
Adjust the Chroma Key properties to get the perfect result.

Creating a Chroma Key Effect with Ulead VideoStudio

Step 1: Load Your Tracks

In VideoStudio, the top video Track in the timeline is called the Video Track and is for standard videos, while the lower video track is called the Overlay Track and is for overlay videos. Place your chroma key track, the interior footage, in the Overlay Track (see Figure 11-5). Put your exterior video in the Video Track.

Step 2: Get a Handle

One nice feature about VideoStudio is that it lets you resize your Overlay Track. You should see handles around it in the preview window (see Figure 11-6). Drag the handles

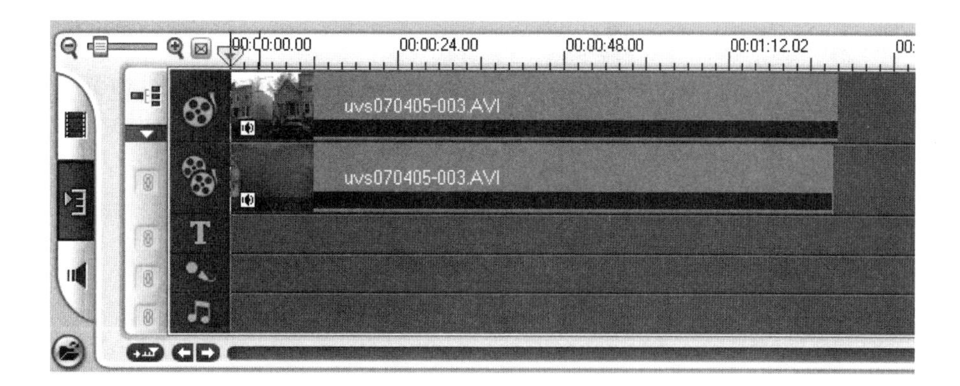

Figure 11-5
Load your footage in the correct tracks. The chroma key footage goes in the Overlay Track.

to get the overlay to the right size, which probably is the full screen for this project. Being able to resize your Overlay Track so easily frees you to do other fun effects in other projects, such as making one actor much, much smaller than another.

Figure 11-6
VideoStudio's resizing handles let you adjust the overlay size.

Step 3: Apply Overlay Attributes

Click the video in the Overlay Track and then click Mask & Chroma Key from the middle of the left side. A new menu will slide into place. On the Attribute tab, check the Apply Overlay Options check box, and then select Chroma Key from the Type drop-down menu if it isn't already selected.

Step 4: Pick Colors

Click the eyedropper tool and use it to select the background color from your Overlay Track. If the results aren't perfect right away, adjust the Similarity setting next to the eyedropper tool (see Figure 11-7). This keys out colors that are similar to the one you picked, even if they're not an exact match.

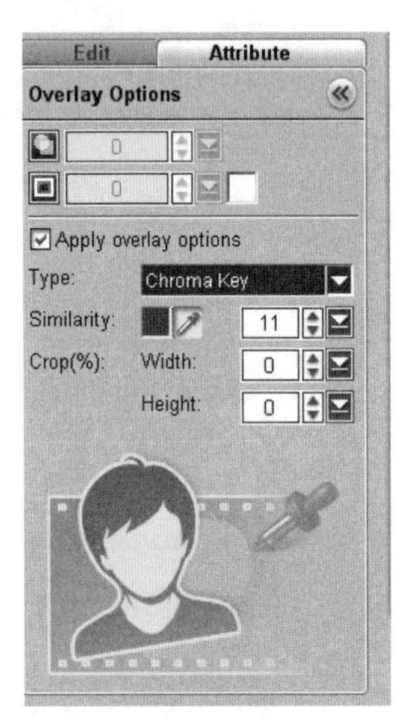

Figure 11-7
Adjust the Similarity setting if the results aren't quite perfect.

There's no end to the fun you can have and the bizarre projects you can create with chroma key editing. Set up a green screen in your house and start making some special effects creations.

Project 12

Create a Dream Sequence

What You'll Need

- **Digital video camera (any)**
- **Home computer**
- **Movie editing software**
- **Cost: $100**

Certainly, most people brandishing video cameras are amateur documentarians at work, documenting the special moments of their lives and recording their children growing up. But some people use the video camera to create, to tell their own stories. If you're capturing your own imagined tales with the camera, there's going to be a time when you want to tell a dream, suggest an altered frame of mind, or just get surreal. If that sounds interesting, then this project is for you.

In this project, we'll explore ways you can create unusual images both when recording and when using editing software. This project shows you how to use lighting to create an unusual mood; how to change colors with your program's color-correction software; how to take advantage of the many filters and effects available in your editing program; and how to use picture-in-picture effects to add a second video to your scene. Dreams take many forms, and if you're going to capture a dream landscape on video, you're going to need to use all the tools available.

Shooting with Colored Lights

Lighting is mood on a movie shoot, so you should always match the lighting you use with the mood you're trying to provoke in your viewers.

A professional shoot uses powerful lights on adjustable stands. The lighting director can change the color of the lights with gels, which are thick plastic sheets in various colors. Gels are inserted into a slot in front of the light to change the color. If you're serious about moviemaking, you might want to invest in professional lights. You can get a single light and some gels starting at around $150, if you visit a professional camera and video shop. Gels are more often sold in bundles, though, which are good deals if you can afford them. Bundles typically include other pieces of equipment that work with the lights.

If you're on a budget, a camera and video shop still has something for you. You can purchase a filter holder, which clamps onto the lamp you're using, for about $40. Pick up a few gels and then use the frame and the gels with a powerful desk lamp.

When you're lighting with a gel, you're simply shooting light through a colored piece of plastic to create light of that color. So, as long as you're careful not to get the plastic too close to the lamp and start a fire, you're free to experiment with different types of instant gels. Translucent, colored shower curtains are a great choice. You can pick them up at discount stores for little money, and they come in a range of colors. Get ones that are clear enough to allow a lot of light to pass through, but that have a tinge of color. Have a friend hold a shower curtain in front of your light source while you're filming, instead of simply (and dangerously) draping it over the light.

An even simpler solution is to skip the gel altogether and use colored light bulbs. Called "party bulbs," these sell for close to the same price as standard light bulbs and come in shades such as blue, red, yellow, and green. You can find them at well-stocked drug stores or hardware stores. You'll even find color-changing bulbs, which shift from one color to another automatically.

Using Software Colorization

Besides using colored lighting, you can also alter your onscreen colors by using the image control features in your video editing program. You're best off doing this with Adobe Premiere Elements, which offers the most fine-tuning, although most editing programs have some type of color correction options. It's advanced tools like this one that make people buy Premiere Elements. (And it's the product's poor manual that makes people buy how-to books.)

Apple iMovie has several visual effects that you can add to modify the color onscreen. They're not as sophisticated as the Premiere Elements controls, but you can use them to create some odd images.

Color Control with Adobe Premiere Elements

Adobe Premiere Elements has excellent tools for altering color on your footage.

Step 1: Set Up Your Video

Load your video into Premiere Elements and drag it into the timeline viewer along the bottom.

Step 2: Choose Image Control Options

The Properties menu sits at the top-right corner of the screen. Find the Image Control option and click the triangle next to it. It opens to display four different values (see Figure 12-1): Brightness, Contrast, Hue, and Saturation.

Figure 12-1

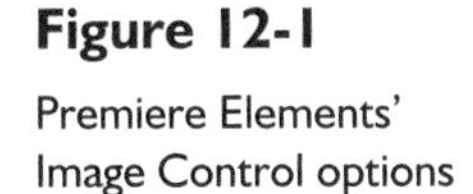

Premiere Elements' Image Control options

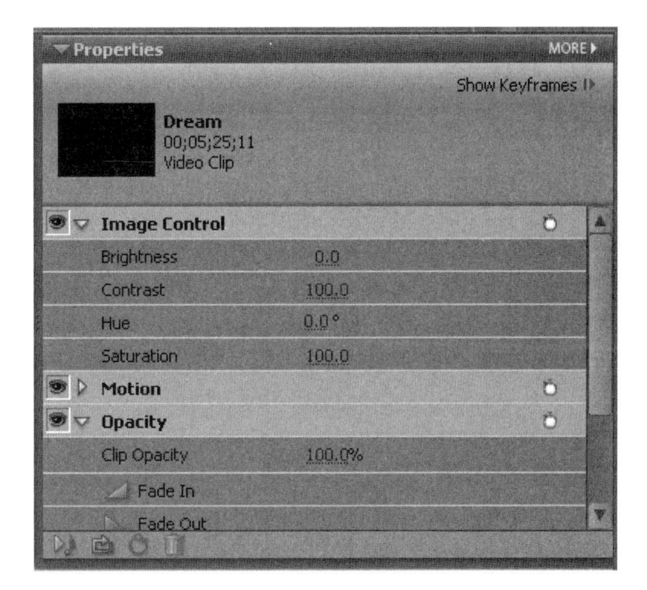

Step 3: Odd Controls

Using Premiere Elements' controls can be a little awkward. You can change the value for any of these four items by clicking its setting and typing a new value in the input box. The better way, though, is to click the value you want to change and then slide your mouse to the left or right. As you do so, you'll see the value increase or decrease. Doing it this way allows you to see how your video changes incrementally, so that you can find the exact spot that produces the look you want.

Step 4: Dream Effects

Changing the Brightness control adds shadow or makes your video look washed out, while changing the Contrast sharpens the lights and darks, or smoothes out the differences. Hue and Saturation are better suited for dream sequences. Changing the Hue enables you to pull out different colors, to give everything a blue or a green cast, for example. Altering the Saturation either drains the color from your work or gives it a hyper-saturated, too vivid color.

tip *You can use these same controls when your video has an unpleasant color that you want to get rid of. For example, you might have white-balanced your camera for fluorescent lights and then forgotten to change it back when you next shot outside. Use the Image Control options to remove distorted colors.*

Step 5: Keyframe Control

Using the Image Control values as just described will change the look of the entire clip, but if you want to gradually add or remove color effects, you need to use keyframes. Don't panic if this sounds difficult; it really isn't. With the Image Control options showing, click the Show Keyframes button in the top-right corner of the Properties menu.

Step 6: Toggle Animation

Next, you'll see a small white circle to the right of the words Image Control shown in Figure 12-1. This is the Toggle Animation button. Click it.

Step 7: Create a Keyframe

You now have a mini timeline to the right of your image values, as shown in Figure 12-2. Creating keyframes, which are simply frames in which an effect begins or ends, is easy. Place the playhead in the timeline at the bottom of the screen where you want your color shift to start. Change the image values in the Properties menu, if necessary. When you do, a keyframe is automatically created.

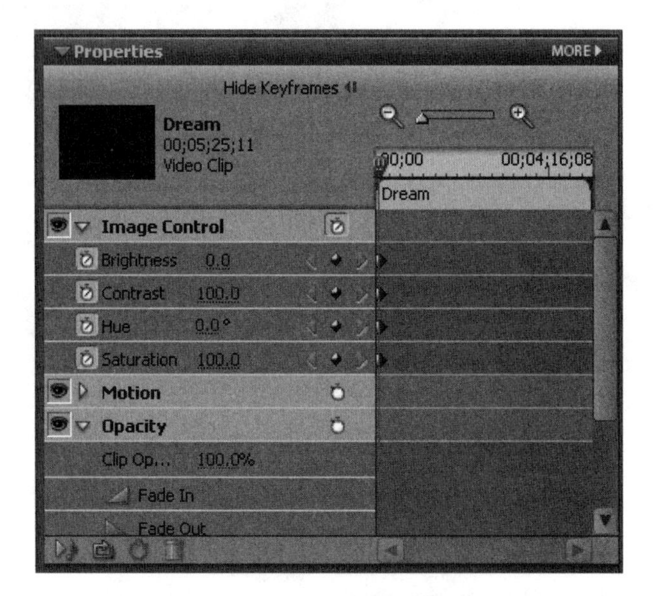

Figure 12-2
Creating keyframes with Premiere Elements

Step 8: Make a Color Shift

Move the playhead to the frame where the color shift should end. Again, change the values in the Properties menu. A new keyframe marker will automatically be created. The software will fill in the effect for the in-between frames, so that the color change will gradually build or decline to match the settings you've given. Preview your clip to see how smoothly it works.

Color Control with Apple iMovie

iMovie doesn't have the same kind of fine-tuning that Premiere Elements offers, but it does have several filters that do similar things.

Step 1: Open the List of Video FX Options

Load your video into iMovie, click the Editing button at the bottom of the window, and then click Video FX at the top.

Step 2: Review the Quartz Options

You'll see a long list of effects that you can add to clips. Scroll down and click the triangle next to Quartz Composer. You'll see more effects, as shown in Figure 12-3, several of which are great for changing color.

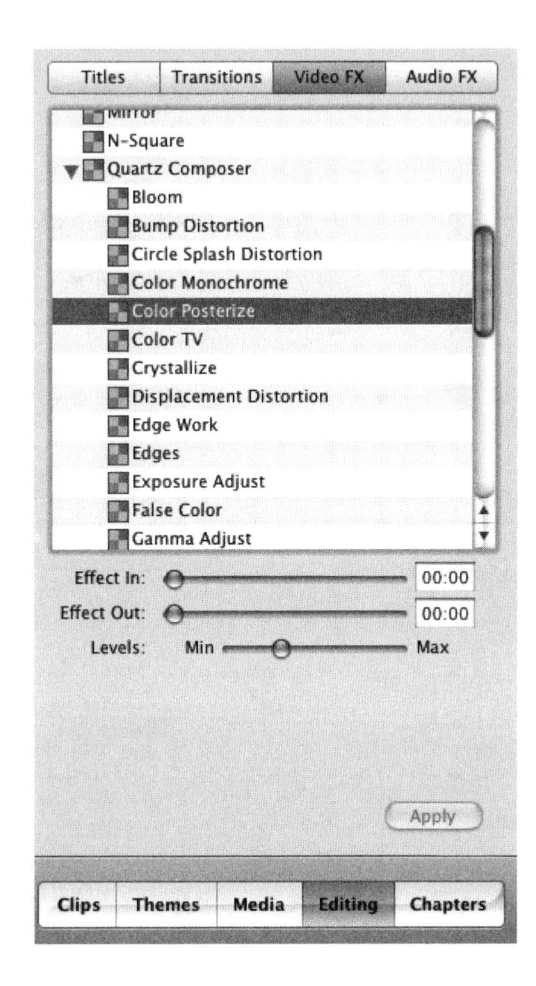

Figure 12-3
Open iMovie's Video FX options.

Step 3: Adjust the Properties

Click any effect, such as False Color, and you'll immediately see a preview of your clip with the effect in the preview window. This is just a preview; you haven't actually added the effect yet. Adjust the effect's controls to get a look that you like. You might have to select a color from a color picker, or use a slider to set the amount of an effect that you want, depending on what effect you chose.

Step 4: Specify Starting and Stopping Points

Under the list of effects, you'll see two sliders that let you set when your effect will start and when it will end, as shown in Figure 12-4. These settings aren't absolutes; iMovie gradually begins and ends effects, so that the change isn't startling.

Figure 12-4

Tell iMovie when to start and stop the effect.

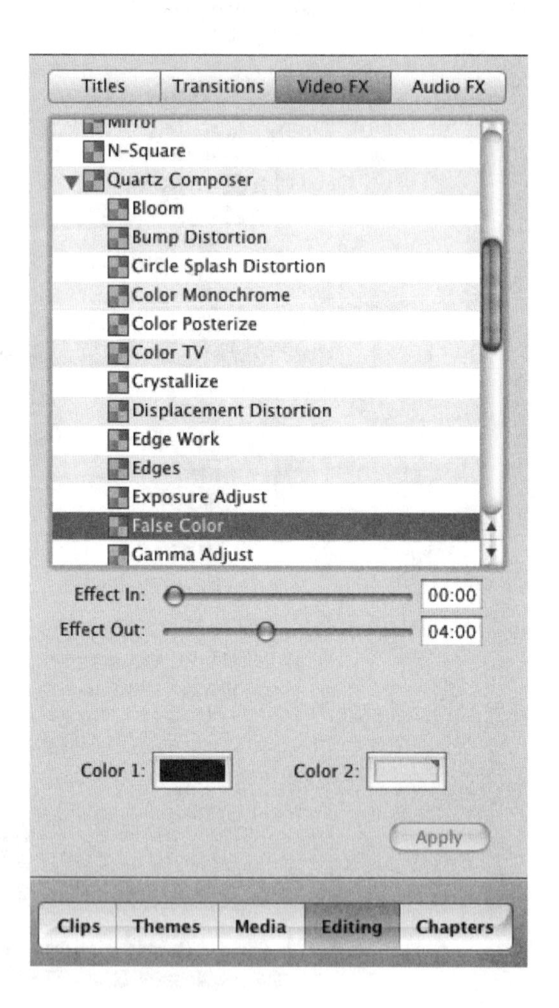

Step 5: Be Sure to Apply the Effect

When you've got the effect that you want, click Apply in the lower-right corner. If you don't click this button, your settings will be lost.

Adding an effect in iMovie can take several minutes. Be patient while your computer renders the video with the effect added.

Adding Picture-in-Picture Effects

Another way to add a surreal wonderland feel to your dream sequence is to add a smaller picture within the larger picture, one that shows some other scene. As you saw in Project 11, Ulead VideoStudio, with its Overlay Track is great for adding second images.

Step 1: Add an Overlay

Select the timeline view along the bottom by clicking the button at the lower left that looks like a bracket. You'll see different tracks along the bottom. Add your main clip into the upper video track. Add your secondary clip, the one that you want to show as a picture inside the picture, in the second video track. This is the Overlay Track.

Step 2: Resize, Reshape

Click once on the Overlay Track to bring up the overlay attributes in the upper-left corner. You'll see your Overlay Track in the preview window, with dotted lines and handles around it. Move it to the part of the screen where you want it to start, and resize it to the correct proportions. You can even drag some of the handles inward, to make the image appear tilted, a cool effect for a dream sequence (see Figure 12-5).

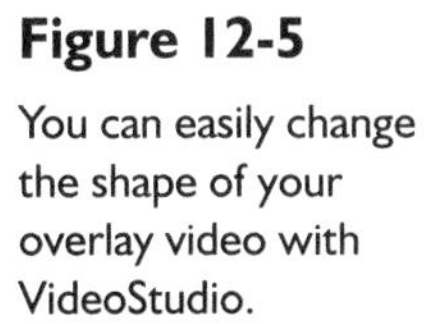

Figure 12-5

You can easily change the shape of your overlay video with VideoStudio.

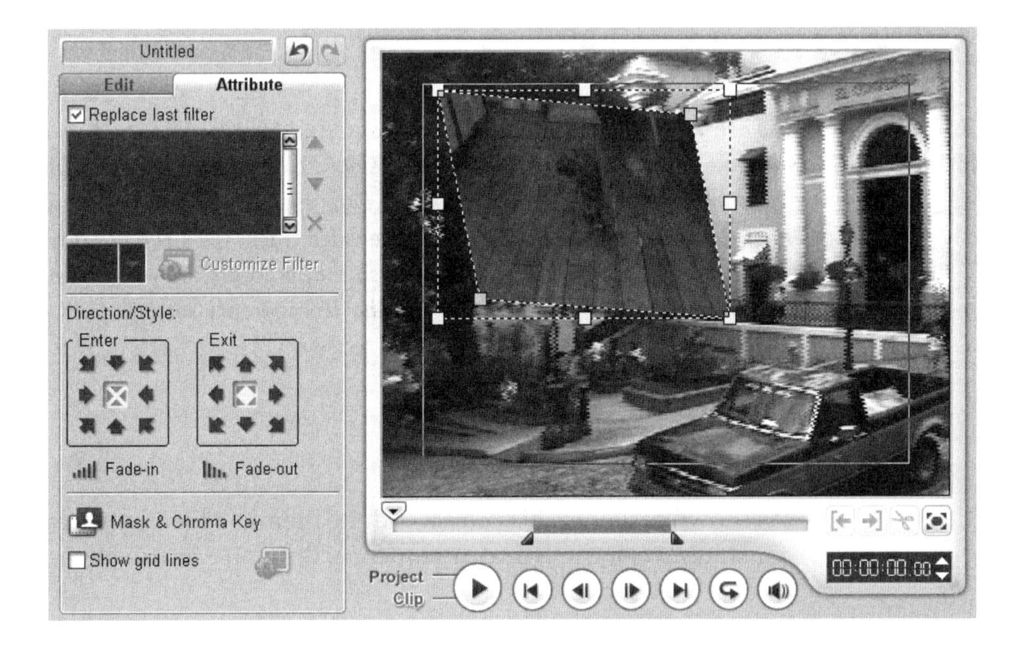

Step 3: Entrance and Exit

You'll see two boxes on the left side with arrows in them. These let you set the direction from which your picture-in-a-picture enters the frame and the direction in which it leaves the frame. In the first box, click the arrow showing the direction from which your overlay image should enter, as shown in Figure 12-6. In the second box, click the arrow showing the direction in which it should leave the screen. Animating your video is that easy.

Figure 12-6

Tell VideoStudio how your overlay video should enter the screen and how it should leave.

tip *You can also use VideoStudio's Overlay Track to easily add a color filter to a scene. Click the drop-down menu in the upper-right corner and select Color. Choose the color filter you want from the list and drag it into your Overlay Track. Drag the clip's right side out, so that it's as long as either the scene or the entire video in the video track. Click the Overlay Track and then resize the color box in the preview window so that it covers the whole screen. Click the Attribute tab in the upper left, and then adjust the Transparency settings until you can see through the color overlay.*

Adding Keyframe Effects

As shown earlier in this project, Adobe Premiere Elements' keyframe tools are useful for shifting color in a dream sequence. But you can use keyframing to add all kinds of psychedelic effects.

Step 1: Access the Distortion Effects

Open the Premiere Elements effects list by clicking the Effects and Transitions button in the upper-left corner. It's the purple icon a little bit down from the top. Click the triangle next to Video Effects, and then click the triangle next to Distort. You'll see several distortion effects that you can select, along with preview images.

Step 2: Add Your Effect

Click-and-drag Wave Warp all the way over to the right side and drop it in the Properties menu (see Figure 12-7). You'll see that Wave Warp now has an entry there. Click the triangle next to its name to expand its options.

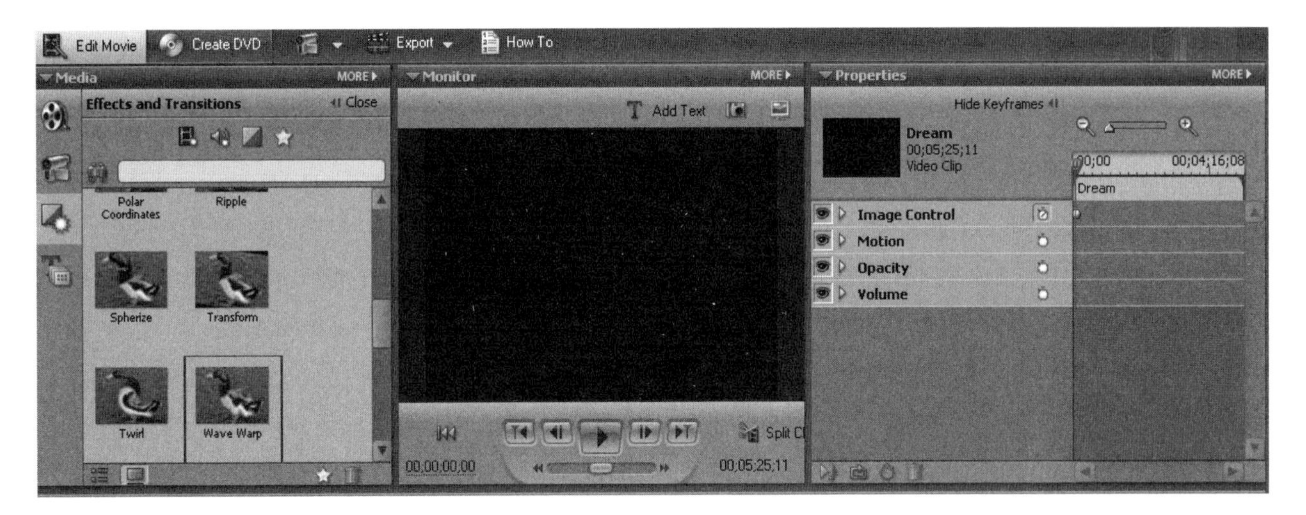

Figure 12-7 To add an effect, drag it to the Properties menu on the right.

Step 3: Create Keyframes

Put your playhead at the first frame of the movie, if it's not there already. In the Properties menu, click Show Keyframes and then click the white circle next to the words Wave Warp. This allows you to animate your effect instead of applying it to the entire clip. Select a starting value for the Wave Warp effect by adjusting the controls in the Properties menu. Move the playhead cursor farther into the clip, and then adjust the properties to something new. To make the effect go away, set the values to zero. The program automatically creates keyframes wherever you change the effect's value, so you've animated the effect just by selecting two frames where the effect should change.

You can use this type of effect animation to introduce all kinds of surreal motion to your movies. Other effects in Premiere Elements let you blur or stylize your video.

Why settle for reality when you can make a magical work out of your video so easily? With the effects found in today's consumer editing titles, you can add a surreal feel with surprisingly little effort.

Project 13

Shooting Sports Events

What You'll Need

- **Digital video camera (any)**
- **Home computer**
- **Movie editing software**
- **Cost: $100**

When it comes to video camera use, shooting sports events probably ranks up there with shooting school plays and choir concerts. Capturing your child on the field, creating an archive that you can dust off later after he or she has won the Heisman or an Olympic gold medal, is just a part of being a good parent these days. Of course, there are other reasons to shoot sporting events: you might want to review the games or matches with your athlete to improve performance, or maybe you want to record your own weekend softball league to make a highlights reel for the end of the season.

Whatever your reason for capturing the action, you have the choice of simply recording the game or making a compelling movie out of it. Dutifully taping the proceedings from your seat in the bleachers will create a record of the event, but it won't make a movie that anyone will want to watch. In this project, you'll learn how to get the best results out of your camera when shooting sports. Then, you'll learn how to shape the results into a compelling movie using standard consumer-level editing software. It'll take a little more work, but you'll get results that you'll be proud to show off.

Adjusting Your Camera

Getting great sports coverage starts with adjusting your camera correctly. Filming sporting events has its own set of challenges: often dim or too-harsh lighting, fast action, and players moving close in and then far away. You need to set your camera so that it's equipped to grab the action and produce the best results.

Sports Mode

All video cameras have a sports mode, which is useful when working with fast movements. Experiment with it to see if you like the results. The final image might appear a little jerky, depending on your camera.

Step 1: Choose the Mode

Turn on your camera and make sure it's not in auto mode. The controls vary from camera to camera, but many have an auto mode that does all the work for you, but doesn't let you change any of the settings. You want to be in program mode. Look for a slider switch on your camera to go into this more advanced mode.

Step 2: Find the Settings

Call up the various shooting settings on your camera. This can be surprisingly tricky to do, because some cameras don't have the controls marked. If your camera has a selection wheel for moving between and choosing onscreen options, turn the camera on and then click the selection wheel. You should see a list of shooting settings appear on the screen. Scroll to the sports mode and select it (see Figure 13-1). Remember to return to auto mode when you're done shooting the event.

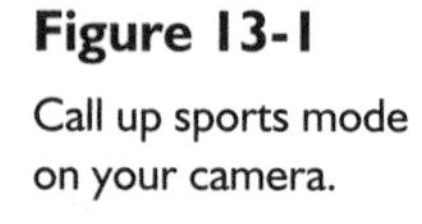

Figure 13-1
Call up sports mode on your camera.

Shutter Speed Adjustments

You also have the option to adjust the shutter speed manually. This might sound difficult if you're an amateur who is unused to changing the exposure, but it's not that hard if you remember some basic numbers:

- If you're shooting outdoor sports with good natural light, you can use a fast shutter speed, such as 1/2000 or 1/4000. This will keep the action crisp.
- If you're shooting an auto race, you need to make your shutter speed a little slower. Try 1/1000, 1/500, or 1/250.
- Indoor sports can be the most challenging of all for your video camera, since the lighting is often dim but the action is fast. Choose a slow shutter speed to compensate for the indoor lighting, such as 1/100.

To find and adjust the shutter speed controls, follow these steps.

Step 1: Enter Program Mode

Turn your camera on and make sure it's in program mode, as described earlier in this section.

Step 2: Choose Menu Options

Click the Menu button and look for a Camera Set Up option. Select it.

Step 3: Set Your Speed

You should see an option for Shutter Speed. Select it. You'll then get several shutter speed options, including Auto and a variety of speeds. Scroll to the one you want and select it, as shown in Figure 13-2.

AE Shift

All cameras have an option called AE Shift, which stands for Auto Exposure Shift. Using this option lets your camera better control how much light enters the lens. You can use it at any time, but it's a strong option for shooting indoor sporting events, since the lighting can vary greatly as you follow the action. It's also a good way to avoid problems from strong backlighting or overexposed scenes.

To activate AE Shift, follow these instructions. The actual controls vary from camera to camera, so yours might not be exactly the same.

Step 1: Enter Program Mode

Turn your camera on and make sure it's in program mode, as described earlier this project.

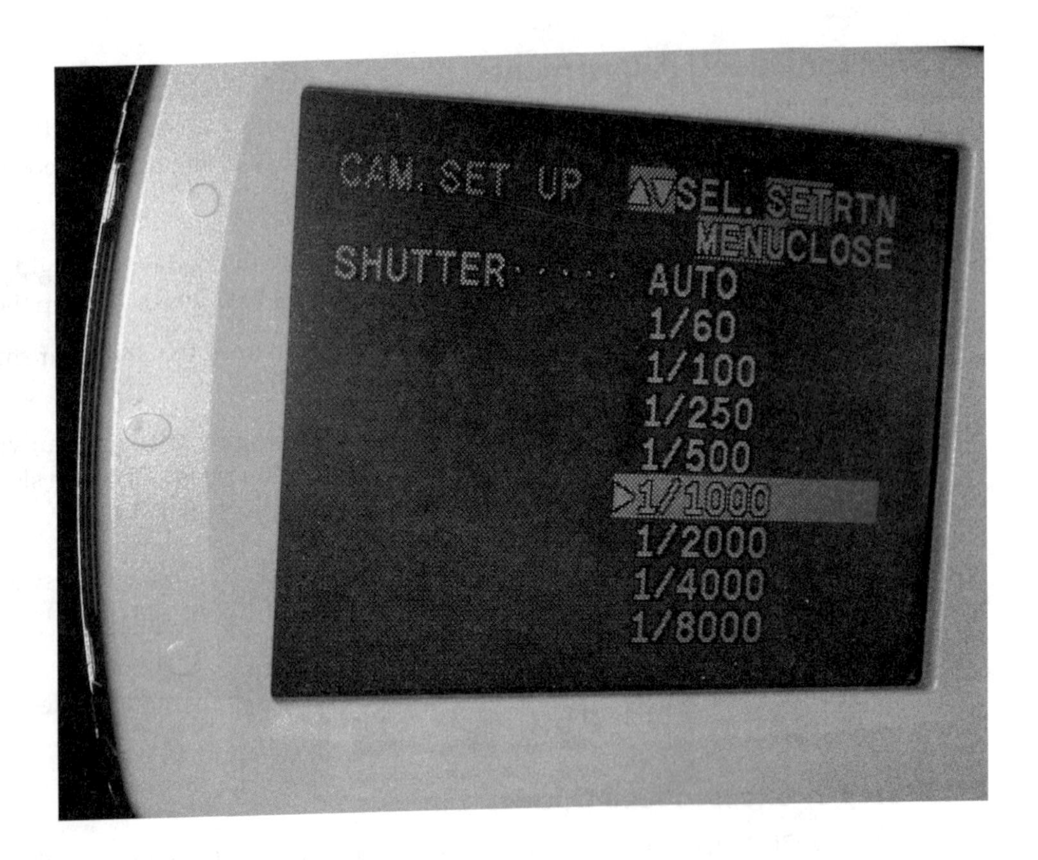

Figure 13-2

Choose the right shutter speed for your sport.

Step 2: Enter Sports Mode

Press the selection wheel to see your list of shooting modes and select the sports mode.

Step 3: Turn on AE Shift

Press the AE Shift button, which is probably located on the side of your camera that has the control buttons (see Figure 13-3). This turns on AE Shift.

Step 4: Make an Adjustment

Now that AE Shift is turned on, you can adjust the exposure setting with the selection wheel. Turn the exposure down, into negative numbers, to allow less light in and dim the image. Turn the exposure up to allow more light in and brighten the image.

Shooting the Action

Once you've adjusted your video camera to optimize it for shooting your sports event, you're ready to film the action. You don't want to just plant your camera and shoot everything from the same position (leave that to the other parents). Follow these steps to get more than just an overview.

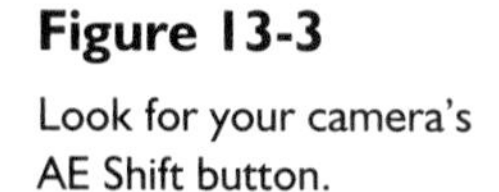

Figure 13-3

Look for your camera's AE Shift button.

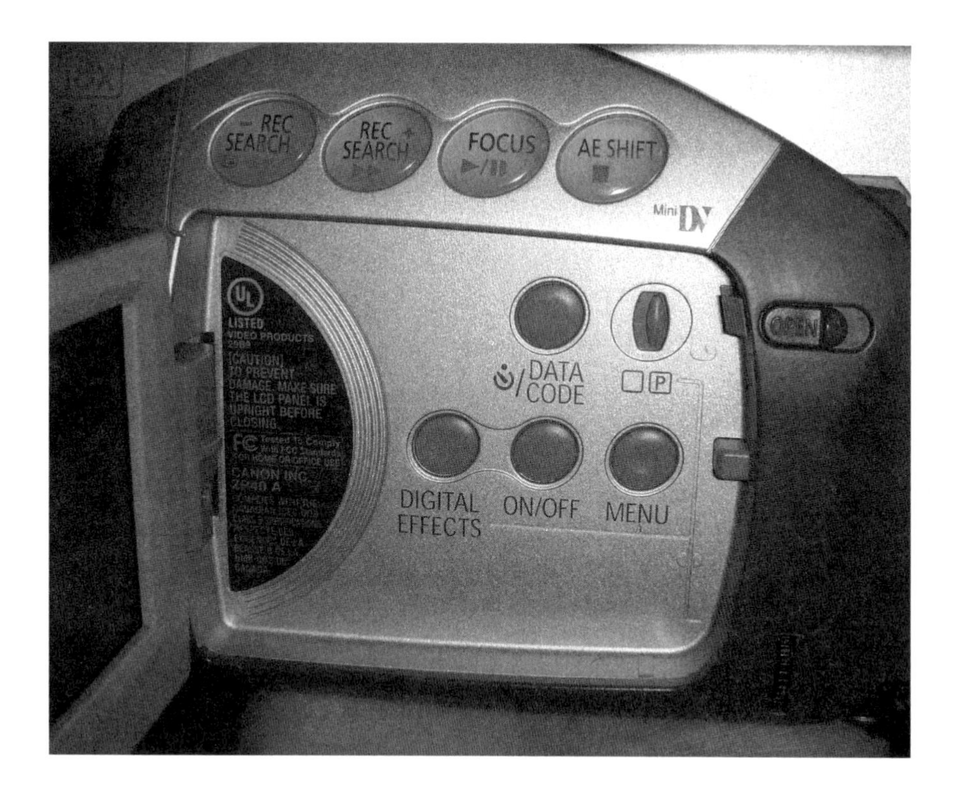

Step 1: Talk It Up

Record a little pregame talk. Sure, you know the specific details about the game now, but will you next year or ten years from now? And what about other people viewing the video? Tape yourself and someone else giving a little pregame chatter. Tell which teams are playing, where they are in the season, and what the expected outcome is. Give a few stats, like the win-loss record for both teams so far, and name the players to watch (starting out with any related to you, naturally).

Step 2: Pick Your Spot

Stake out a good position before the game, and that doesn't mean in the bleachers. Find a spot closer to the action, but one where you're not blocking anyone else's view. Be a good citizen and ask the coach's permission if you suspect you might be in the way.

Step 3: Change Your Viewing Angle

Change your viewing angle, if possible, frequently during the game. Sports videos are more interesting when there are several different angles on the action. If the area permits, follow the action physically instead of relying on your camera's zoom.

Step 4: Avoid Zooming

Speaking of zoom, try not to use it much. You'll be tempted to zoom in on the players at the center of the action, especially if one of them is your kid, but then the action will shift and you'll be jerking your camera into place trying to follow it. You're better off relying on wide shots to cover a larger area of the court or field. Get in close to the action by standing as close to the players as possible, not by zooming.

Step 5: Get the Right Slant

Tilt your camera for exciting action shots, but do so sparingly. If you can see that someone is coming up for a basket and you're under the net, for example, tilting your video camera can make the shot more dramatic.

Step 6: Mum's the Word

Stay silent while you record. You might feel like the worst parent in the world if you do not cheer for your kid, but staying silent will result in much better video. When someone who is holding a camera cheers while recording an event, their voice overwhelms any surrounding sound. The result is distracting to anyone who is watching the video.

Step 7: Get B-Roll Footage

Get some B-roll footage to mix in with the action shots. Shoot the crowd cheering once in a while, or shoot the cheerleaders on the sidelines. Capture some of the band's halftime show. Also, it's a good idea to take a shot of the scoreboard after every score. This makes it much easier for anyone watching the video to follow along.

Those suggestions will take you through a day of shooting and give you great footage. Here are a few other tips for shooting sports:

- Watch what's going on around you. Sometimes the action leaves the field, and if you're standing nearby, you could get creamed. Keep one eye on your camera and the other on the action around you.
- If you're coaching a game or a practice session and you want to use the footage you record as a training tool, look for a camera that has slow-motion playback built in. You can then record a play and instantly show the results to your players, slowing the action down to better illustrate the point you're making.
- Shooting sports is one of the few times when you're better off not using a tripod. Sure, a tripod will steady the image, but it also slows you down when you're trying to move the camera or change the shooting angle. If you need to brace your camera, consider buying a monopod, which is a telescoping pole that you can use to steady your video camera. Monopods fold up small and start at around $20.

- If you're simply holding your camera without a tripod or monopod, hold it as steadily as possible. Use one arm to hold the camera and the other to brace the first arm by pinning it tightly against your body. If you need to zoom in on something, you'll need to find something to rest your camera on, since there's no way you can keep your camera steady while zooming in.
- If you want to capture swimming or diving, check out the waterproof video camera cases that are available. The online store Waterproof Cases.net (www.waterproofcases.net) is a good place to start. Obviously, you can't record the action underwater during a swimming and diving meet, but you can shoot some interesting footage during practice sessions, if you're willing to get wet.

Editing Your Footage

You're not done with your movie when you're done shooting (at least, you shouldn't be). You need to edit out the dull spots to tighten it up, and to intersperse some of that B-roll footage to make the video more interesting. Here are few other ideas for making better sports videos.

Slow-Motion Replays

When you've captured a really great scene, do like the professional sports programs do and replay it in slo-mo so that your viewers can savor the action.

Step 1: Cut It Out

In Adobe Premiere Elements, cut around the moment you want to repeat so that you can isolate just that play. Put the playhead at the start of the play in the timeline and click the Splice Clip button below the Preview window. Then, put the playhead at the end of the play and click Splice Clip again.

Step 2: Make a Copy

Click your newly spliced scene. Choose Edit | Copy. Then, choose Edit | Paste. You'll see a copy of your clip added in an unused video track of the timeline, just above or below the original (see Figure 13-4).

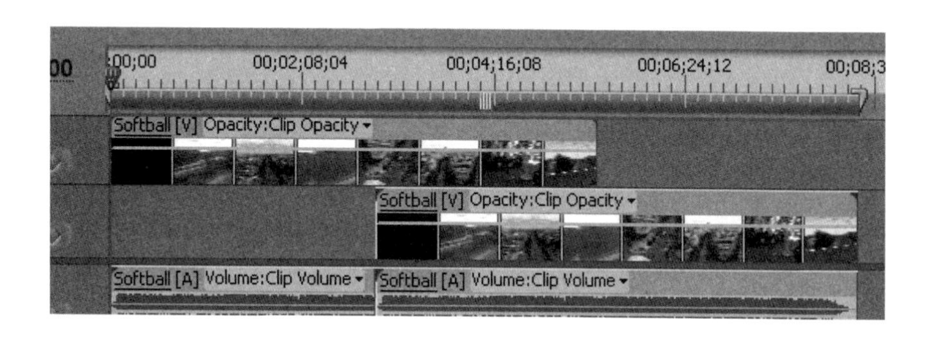

Figure 13-4 The copy of your video clip will appear just above or below the original.

Step 3: Stretch It

Click the Time Stretch button, which is just above the timeline and to the left. It has a circular button with a wavy line through it, as shown in Figure 13-5.

Figure 13-5

The Time Stretch button

Step 4: Such a Drag

Drag the right edge of your copied clip to the right to lengthen the scene and slow it down (see Figure 13-6). A pop-up window will let you know how much time you're adding. Preview the results. You can always shorten or lengthen the clip again to get the right results.

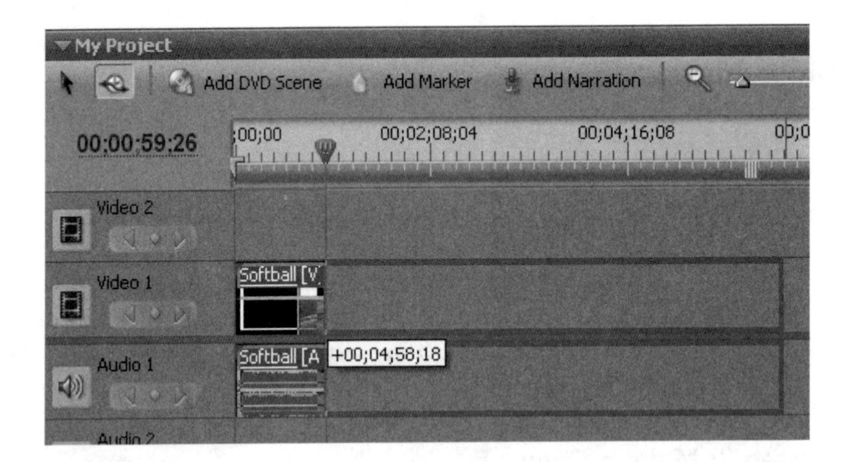

Figure 13-6

You can create a slow-motion effect just by lengthening a clip.

Step 5: Make Room for the Slow-Motion Scene

Click the Selection Tool button (to the left of the Time Stretch button) to get out of Time Stretch mode. In the original video track, drag after your cut the footage located to the right to make room for the slow-motion scene. Then, drag your slow-motion clip into the original track.

tip *If you're going for comedy, you can use the same Time Stretch tool to compress a clip, thereby putting it into fast motion.*

Add Narration

The time to narrate the game isn't when you're recording it, but rather when you're creating the movie. Follow these steps to add your own play-by-play and commentary.

Step 1: Mic Your Computer

Connect a microphone to your computer, if necessary. Many have built inmics, but not all computers do.

Step 2: Pick Your Moment

Place the playhead in the timeline at the moment where you'd like to begin your narration.

Step 3: Talk Time

Click the Add Narration button. In Premiere Elements, this is located just above the timeline. A small window with recording controls opens (see Figure 13-7). Click the red recording button to start recording. The movie will play as you record.

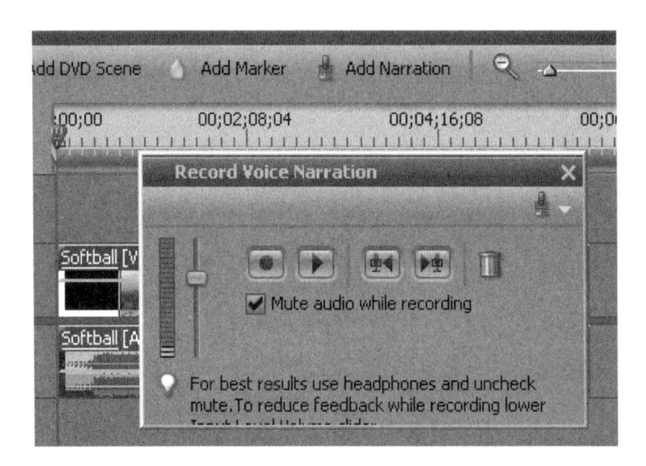

Figure 13-7 Adobe Premiere Elements' narration controls

With a little effort, your sports videos will have all the excitement of sports on television, with the advantage that you know the players. Take a few moments to prepare your camera before the sports event, and then edit your footage afterwards; you'll be rewarded with videos that are just as exciting as the real events.

Project 14
Create an Interactive DVD Game

What You'll Need

- **Digital video camera (any)**
- **Home computer**
- **Movie editing software**
- **Cost: $100**

This project will guide you through the process of creating an interactive DVD game, which people can then play with a DVD player. This is unlike the other projects in this book, and it's probably the hardest to create. While other projects helped you create traditional movies on different themes, or gave you techniques for using your video camera or editing software more successfully, this project is all about planning. You need to thoroughly plan out a DVD game before you start filming it, and stick to a strict labeling system all the way through. If you don't, the end DVD just won't work correctly.

You'll actually learn how to create two different DVD games with this project, but both work in the same way. In one, viewers will select answers to multiple-choice questions given on the screen. In the second, viewers will be guided through a choose-your-own-adventure-style story. Both projects work by asking viewers to select an option, which then determines the video clip that they'll view next. As you work, you'll get to know a lot about creating a detailed menu structure on a DVD.

About Interactive DVDs

You've probably seen interactive DVD games in stores, or perhaps you even own one. Games such as *Scene It* are played through a DVD player and let players answer questions and view video clips. Unfortunately, it's not currently possible to make something as involved as the commercial games with consumer video editing and

DVD-creation software. You can't put selection buttons within a video that cause another clip to play, and you can't keep score of results. Nonetheless, you can still make a fun game within the limits of the DVD menu structure. It won't have the same level of polish, but it will still be a fun game for your children to play with their friends (or to create with their friends).

You're going to create clips that ask people to select a response to something on the screen. They do this by playing a different clip, one that the first clip guides them to. The first project is a simple quiz. You can select the subject and make the questions; the steps below will show you how to set it up so that viewers can choose any of three possible answers to a multiple-choice question. The second project is an adventure story that enables the viewer to decide how the action should proceed by choosing between two completely different responses. These steps will guide you through both projects simultaneously, since they both work the same way.

Planning the Quiz

As previously stated, the key to both of these projects is planning. You need to know exactly what you're going to present before you start filming, so get a pad of paper and follow these steps.

Step 1: Pick Your Topic

In this project, you're going to create a DVD game with five multiple-choice questions. Each segment will be filmed, so you'll need to ask the questions on camera, or present them through some kind of scene. Your first step, then, is simply to decide what type of quiz you'd like to create. This is a great project for kids to do as a school project, perhaps for extra credit, which would mean basing the questions on recent school lessons. You could also create questions on any topic that interests you, from items in the news to pop culture trivia. If you want to have people play the quiz at a family gathering, base the questions on family history.

Step 2: Write Down Your Questions and Answers

Write out your five multiple-choice questions and the three answer choices for each. You'll need to record a segment for each question and each response. Since the DVD will have five questions, you'll be creating 20 video segments total. Think about how you'd like to present the questions and answers—what type of theme you want to give the project. You could present them from behind a desk, at the beach, or anywhere else. After you've written down your multiple-choice questions and the corresponding answer choices, number them 1-Q, 1-A, 1-B, 1-C, 2-Q, 2-A, 2-B, 2-C, and so on. (Be sure the correct answer isn't the same option for each question.) You'll use these codes later on when you are recording and editing, where they'll keep you from getting lost.

Planning the DVD Adventure

The DVD adventure doesn't have the same tidy organization as the quiz, so you'll need to plan it out especially well before you begin.

The idea is to present a story in which the viewer is given the choice between two scenes at various points in the story, so that the viewer can shape the direction of the story. Each scene ends with the choice between two more scenes, so the story continues to branch off in different directions based on which scenes the viewer chooses.

Step 1: Choose Your Story

This project puts you in the role of storyteller, and, just like the quiz project, you're free to choose any subject you like. The difference is that the quiz is merely about asking questions, whereas this project requires you to tell a story. It should be an adventure story in which the viewer tries to guide the characters to a happy ending. Your first step is simply to decide what type of adventure story you'd like to tell.

Step 2: Create the Story Tree

Write out your storyline on a big sheet of paper, with the different options branching off like the branches of a tree diagram (see Figure 14-1). Here, your tree shape will go sideways, instead of up and down. So if your first clip ends with the characters choosing either the left door or the right door, your diagram will show two branches off that first idea. Each branch represents a different option in the story. Continue each branch

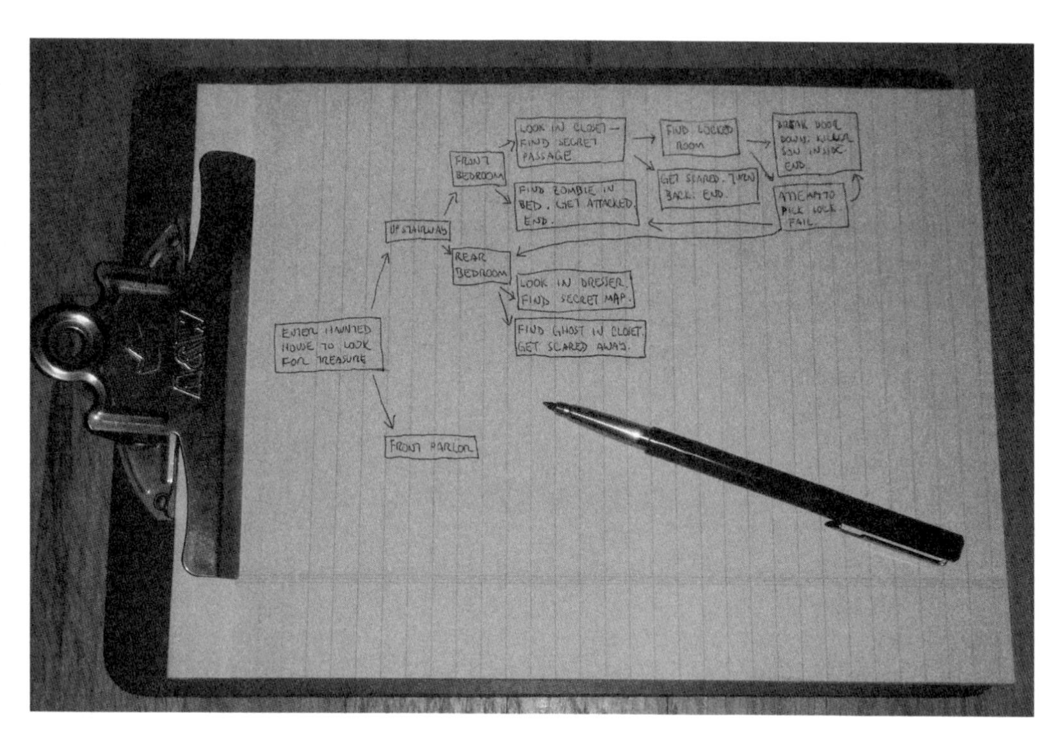

Figure 14-1

Write out your story in a sideways tree diagram.

with two more options, furthering each possible story. The branches don't have to all be the same length. In fact, it's better if bad choices lead to an early end (perhaps a grisly demise for the characters). The way we're laying out this exercise, you'll need 51 clips total (each segment of the story is a new clip). That's a lot to shoot and edit, so if it feels overwhelming, you might want to make a simpler DVD, although then there would be fewer choices for your viewers.

Step 3: Create the Story Grid

Now you need to mix those segments up. Your viewers can't just flip through the DVD in order; that would take away the element of surprise. Get a large sheet of paper and make five columns, labeled A, B, C, D, and E. Write the numbers 1 through 10 under each column, as shown in Figure 14-2. Now, write the various story segments one by one at random in the 50 cells you just made. Be sure to place them randomly. Don't include the first segment, though; that will go in a special area. What you're creating now is a plan for the DVD's menu structure. As you write down a segment in the list, write the codes for the branches that come off of it immediately afterwards, so you know where each segment leads. You'll use these codes later in editing, when you write the story choice titles that your viewers will use.

Figure 14-2
Enter your story segments randomly in a grid.

Record Your Video Segments

After you've got the planning nailed down, you're ready to start filming your segments. You can record both of these projects any way you want: you can act them out with friends, make it a family project, or use toys for your actors. You can film it at your home or anywhere you like.

When filming the quiz, you'll need to shoot the following:

- Five clips that present the five quiz questions and the three possible answers to each
- Five clips that confirm that the viewer picked correctly
- Ten clips that inform the viewer they picked incorrectly and explain why that answer is wrong

After each question, direct the viewer on what to do next by saying, "If you think *x* is correct, click video A, if you think *y* is correct, click video B, and if you think *z* is correct, click video C." Follow the plan you made that spells out all the options.

When filming the adventure story, you don't need to give the next video options verbally, since you'll present them onscreen as text instead. Before you record each segment, say its place in the menu into the camera ("This is A4," for example). That will help you keep track of the many clips later during editing. If you want to be really professional, you can make a scene marker to hold up before each shot, to let yourself know which segment it is. Check off your segments as you shoot them, to make sure you get them all. Keep each clip short, around a minute, to ensure that the story moves quickly and that you don't run out of room on the DVD.

Editing the Quiz and Story

Editing the quiz shouldn't be too demanding, because the segments are all brief and you don't need to do anything special with titles. The important thing to remember is to save the clips by the codes that you used before (1-Q, 1-A, 1-B, 1-C, and so on) so that you can keep them organized. Save them to the same folder on your desktop, so that you can find them easily.

When editing the adventure story, you need to put in titles at the end of each clip to give viewers the correct options for where to go next (see Figure 14-3). You might write, for example, "If you think they should return to the camp, choose video E7 next. If you think they should investigate the tomb, choose A3 next." You could put these titles over a simple black background, but it would look nicer to end each clip with a freeze frame effect with the instructions over the frozen frame (for instructions on freeze frames, see Project 10). When each clip ends, the DVD menu will appear, enabling your viewers to click whichever segment they want to see next. When you save each clip, use the code from your planning sheet (A1, A2, and so on). Save all the clips to the same folder, so that you can find them easily.

Figure 14-3
Put instructions on choosing the next clip at the end of each story segment.

Creating the DVD Menus

Your viewers will access the quiz or story via the DVD menu of each, so take care when constructing it. Choose basic DVD templates, ones with text buttons instead of picture buttons that show a part of the video. After all, you don't want to give away the surprises that are to come, so it's better to stick with text buttons. Using text-only buttons also allows you to fit more options on a page, which will be important when making the adventure story menu.

Creating the Quiz Menu

Step 1: Choose a Simple Template

Open your DVD-creation program or the DVD part of your video editing program. Choose a simple DVD menu template, one with simple text buttons only and no photo buttons.

Step 2: Create the Menu Structure

Put the title of your quiz on the template. Add five subfolders, labeled Question 1 through Question 5 (see Figure 14-4). Don't put any videos on this main page, just the five subfolders.

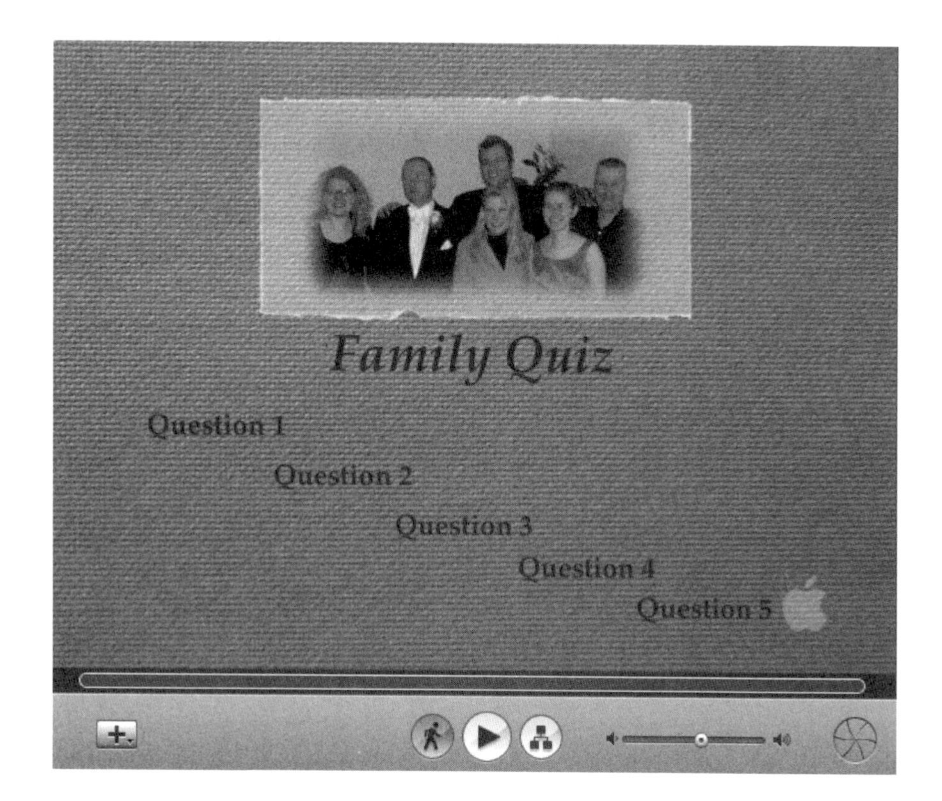

Figure 14-4

The quiz menu structure

Step 3: Place the Buttons

Put one question and the three corresponding answers in the first subfolder (see Figure 14-5). If the subfolder menu template creates photo buttons for your videos, turn this option off. For example, in iDVD, click the Buttons tab at the lower right. Click one of the photo buttons and select the "T," for "text," in the upper-right button menu. This turns your image button into a text-only button. Your program might try to snap your buttons to an ordered grid, but you can turn off this option. In iDVD, click Free Positioning in the Buttons controls. Label each video clearly, so that viewers can see which is the question in each folder and which are the three answers.

Step 4: Populate the Subfolders

Repeat Step 3 for the other four subfolders.

Figure 14-5
A quiz menu subfolder

Step 5: Go for the Burn

Burn your DVD when you're done.

Creating the Adventure DVD Menu

The adventure story DVD menu is similar to the quiz menu, but larger. Creating it should be a snap if you've labeled all of your files correctly. Have your planning sheets with you as you work.

Step 1: Choose a Template

Open your DVD-creation program or the DVD part of your video editing program. Choose a simple DVD menu template, one with simple text buttons only and no photo buttons.

Step 2: Create the Main Menu

Put the title of your adventure story on the template. Put the introductory clip on this first template, labeled something like "The Adventure Begins." Add five subfolders, labeled A, B, C, D, and E. Your main menu should resemble Figure 14-6.

Figure 14-6

The adventure story main menu

Step 3: Fill the Subfolders

Put the correct ten video clips into each folder, according to your planning sheet. Keep the simple labels you used when you saved them (A1, A2, and so on, as shown in Figure 14-7). If the subfolder menu template creates photo buttons for your videos, turn this option off. For example, in iDVD, click the Buttons tab at the lower right. Click one of the photo buttons and select the "T," for "text," in the upper-right button menu. This turns your image button into a text-only button. Your program might try to snap your buttons to an ordered grid, but you can turn off this option. In iDVD, click Free Positioning in the Buttons controls.

Figure 14-7 A subfolder for the adventure story DVD

Step 4: Burn That Disc

Burn your DVD when you're done.

Congratulations! You've just made an interactive DVD that your friends and family can all enjoy. Go on to make others, trying new things each time.

Project 15

Create a Time-Lapse Video

What You'll Need

- **Digital video camera (any)**
- **Tripod**
- **Home computer**
- **Movie editing software**
- **Cost: $200**

It sounds both impossibly difficult and impossibly cool: creating your own time-lapse movie. Cool, certainly, but it's surprisingly easy to achieve. You can create a time-lapse movie of any subject you like by using tools that could be included in your video camera or that come with a few of the better consumer-level video editing programs. This project will guide you through the process, explaining how to set up the shot and offering tips on recording, lighting, and creating a smoothly flowing movie.

Once you're ready to try out time-lapse video, there are lots of subjects all around you that you can choose to apply it to. Point a camera out the window and create a time-lapse video of the clouds going by. Or buy a bunch of flowers still in bud and film them slowly opening. You could document a long process, such as building an addition on your house, turning a weeks-long event into a two-minute movie. One fun idea, though, is to perch your video camera up high and make a movie of a day in the life of your own household.

Creating Time-Lapse Movies with a Camera

The easiest way to create a time-lapse movie is to set your camera to do the work for you. We're saying "camera" here, and not "video camera," because several digital cameras now have this ability.

To see if your camera does, look for a feature called interval shooting. The name means that you can set your camera to shoot at a set interval. While many cameras offer it, many others don't, so you might need to use one of the software solutions described later in this project.

Press your camera's menu button and look for interval shooting. It should be listed under shooting options. When you select it, your camera will ask you to set the interval between shots. Think about what type of event you'll be capturing. If you're shooting clouds moving across the sky, you might want to try every second. If you're shooting a flower opening, perhaps every minute would work well. If you're shooting an event that lasts several days, go even longer. Experiment to find what works best.

tip *Your movie software will create a finished movie at 30 frames per second, so you'll need 30 shots just to make one second of video. You can use this to plan the interval you should use. For example, if you want to end up with a two-minute movie, you'll need 3600 shots (120 seconds × 30 frames each second). So if you're shooting for four hours, you should shoot at an interval of one shot every four seconds (14,400 seconds in four hours divided by 3600 shots).*

Your movie will look better if your video camera grabs only one frame at each interval. Some cameras require you to grab a few frames each time. If this is the case for you, consider using one of the software solutions instead.

tip *Shooting a high number of interval shots takes quite a bit of room. If you're using a standard digital camera, make sure you have a huge memory card first.*

Creating a Time-Lapse Movie with Software

Both Apple iMovie HD and Adobe Premiere Elements let you create time-lapse effects on your computer, instead of on the camera. They give you the option of taking existing footage and speeding it up, to give the impression of a time-lapse movie, or of connecting your video camera to your computer and using the software to drive the camera. Connecting your video camera produces more attractive results, but it limits you to shooting subjects near your desktop computer or dragging your notebook computer with you while you shoot.

Speeding Up a Video with iMovie

This option, speeding up how fast a movie plays, works with older versions of iMovie.

Step 1: Get Ready, Get Set, FX

Open iMovie and load your video into the timeline. Click the Editing button at the bottom of the window, and then click Video FX at the top.

Step 2: Speed It Up

Scroll down until you see Fast/Slow/Reverse, as shown in Figure 15-1, and select it. Then, drag the Speed slider all the way to the left, to make the clip as fast as possible.

Figure 15-1

The speed control is a video effect in iMovie.

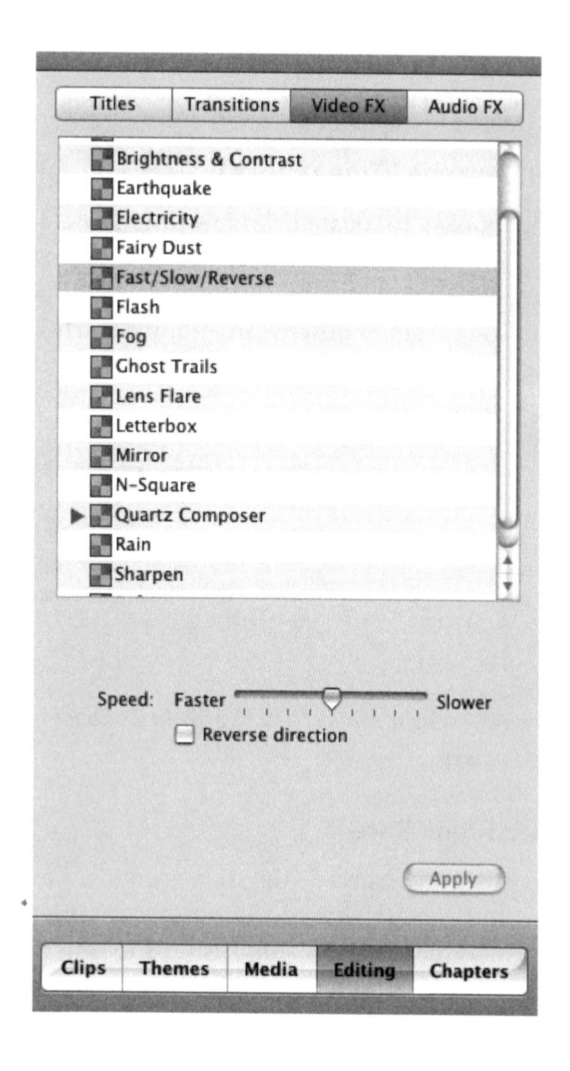

Step 3: And Repeat

If the speeded-up clip isn't fast enough for you, you still have an option. Export the video to your desktop and then re-import it. You can now apply the effect all over again, making the clip even faster.

Your audio will speed up as well when you shorten the length of a video, and the result sounds odd. Consider removing the audio track from your clip before you begin.

Shooting a Time-Lapse Video with iMovie

The ability to create time-lapse movies was added to iMovie with version 6. If you have an earlier version, you'll need to upgrade before you can create one.

Step 1: Get Connected

Start by plugging your video camera into your computer and putting it into Record mode. If you're using a MiniDV camera, remove the tape first so that it doesn't save your work onto the tape. You'll want to record this session directly to your hard drive.

Don't expect your battery to last for an entire time-lapse photography session. Plug in your camera's power cord before you start.

Step 2: Finding the Time (Lapse)

It's not easy to locate iMovie's time-lapse settings. Find the camera-shaped button in the lower-left corner. This puts the program into camera mode, which should have happened automatically when you connected the camera and turned it on. You'll see a small downward-pointing arrow to the left of it. Click it and select Time Lapse (see Figure 15-2).

Figure 15-2
iMovie's almost hidden time-lapse option

tip *If you have an Apple iSight camera, either attached or built in, you can use that to capture time-lapse footage, as well.*

Step 3: The Right Rate

Surprisingly, iMovie's time-lapse controls are incredibly awkward and even require you to do math. Instead of simply asking you the time interval you'd like to use for shooting, iMovie requires you to enter the interval in video frames, as shown in Figure 15-3. Remember that one second of video has 30 frames. So, if you have iMovie save one frame out of every 1800 frames, you'll get one frame every minute (30 frames per second × 60 seconds). Do that for 30 minutes and you'll have one second of video. Set iMovie for whatever frame interval you'd prefer.

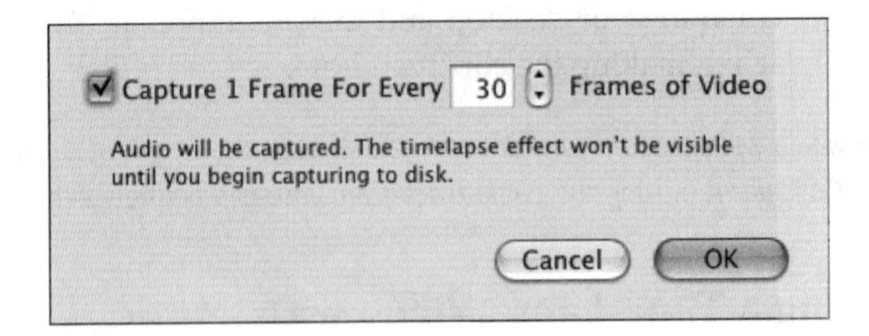

Figure 15-3
Setting iMovie for time-lapse shooting requires a little math.

Step 4: Start It Up

Click the Import button in the middle of the preview window. iMovie will begin saving clips.

Make sure your camera is well braced when shooting time-lapse video, because even the slightest movement could ruin the effect. If you're going to experiment with time-lapse effects, pick up a tripod. It will keep your camera steady and let you shoot in more places.

Step 5: End Times

Unfortunately, iMovie lacks a timer, so you'll need to stop it manually. When you've recorded enough, click the Import button again.

Step 6: Add a Clip

Your clip will show up in the Clips library in the upper-right corner. If the library isn't visible, click the Clips button in the lower-right corner. You can preview a clip by selecting it and clicking the Play button in the center.

You can have iMovie save your time-lapse footage directly to the timeline rather than the clip library, saving you a step. Choose iMovie HD | Preferences. Select the Import options, and select the option to save clips to the movie timeline.

Step 7: Trash the Sound

The audio in a time-lapse movie sounds clipped and staticky, so it's best to remove it. Drag your clip to the timeline. With the clip highlighted in blue, choose Advanced | Extract Audio (see Figure 15-4). This gives you a separate audio track. Drag the audio track to the trash can in the lower-right corner to throw it away.

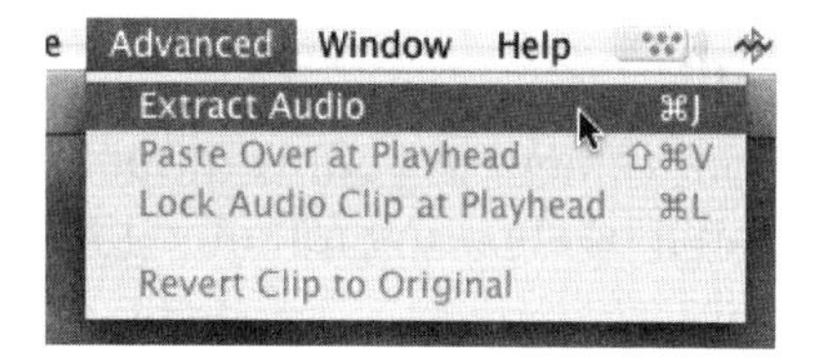

Figure 15-4
Extract the audio from a time-lapse movie, then trash it.

Mac users who want more control over their time-lapse movies should download iStopMotion, a terrific program for making time-lapse or stop-motion videos ($39.95 from www.istopmotion.com).

Speeding Up a Video with Premiere Elements

If you don't have interval mode on your camera and you want to create a time-lapse movie away from your computer, you can still do so. Just shoot as much footage as your camera will allow, and then use Premiere Elements' Time Stretch tool.

Step 1: Stretching—and Shrinking—Time

Load your footage into Premiere Elements and drag in into the timeline. Click the Time Stretch tool just above and to the left of the timeline (it looks like a white circle with a wavy line through it, as shown in Figure 15-5) to enter time stretch mode.

Figure 15-5
Use the Time Stretch tool in Premiere Elements to create fast motion.

Step 2: Drag It

Click the right edge of your movie in the timeline. You'll notice that your cursor changes to a bracket shape. Drag the right edge to the left to shorten the clip, speeding it up and creating a time-lapse effect.

Step 3: Take a Peek

Preview your creation in the center window. Click the pointer icon next to the Time Stretch mode button when you're done, to return to normal editing mode.

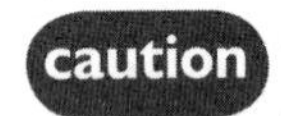
Keep the light source even when recording a time-lapse movie, or you'll end up with an irritating strobe effect.

Shooting a Time-Lapse Video with Premiere Elements

Premiere Elements' controls are a little easier than those of iMovie, so capturing a time-lapse video with an attached camera is pleasantly simple.

Step 1: Power Up

Attach your camera to your computer and put it in record mode. Plug in the power cord if you'll be shooting for a long time. You don't want your battery giving out in the middle.

Step 2: Get the Goods

You need to tell the software that you're grabbing video from a camera. Click the video camera–shaped icon in the upper-left corner, and then choose to import from a DV camera.

Step 3: Save Right

The video import window will open. Click the button that says More in the upper-right corner. Make sure Capture to Timeline is checked. This puts your captured frames in the timeline automatically.

You can preview your Premiere Elements video even during the recording process. That's handy for when you aren't sure that you've gotten the interval right, and you want to make sure you're getting a good effect. Click the Preview button in the lower-right corner of the Capture window at any time to see what you've got so far.

Step 4: Time-Lapse Controls

At the top center, you'll see a button that says Stop Motion. Click this button, and then choose Create New Stop Motion from the next window (see Figure 15-6). Don't worry that it doesn't say "time lapse;" the same tools are used to create time-lapse and stop-motion effects. You should now see a preview image in your center window.

Figure 15-6 Premiere Elements' time-lapse and stop-motion controls

Step 5: Hold the Onions

Onion Skinning will already be checked in the lower-right corner. Uncheck it. That's a tool for stop-motion projects that allows you to see a shadow image of one or more previous frames over the current one.

Step 6: Setting the Time

Click the Time Lapse button in the lower left, and then click Set Time right next to it. This opens a window, shown in Figure 15-7, which lets you set the time interval between frame captures and the length of time you'd like to record. Set both values, and then click OK.

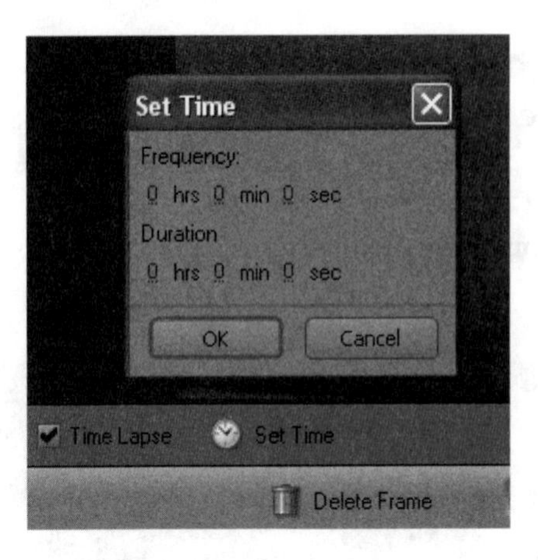

Figure 15-7
Setting the values for interval recording

Step 7: Start Recording

Click the Start Time Lapse button in the bottom center. Your recording session will begin.

Step 8: And Save

When finished, close the Capture window. A pop-up window will ask if you'd like to save your results as a movie file. Click to do so.

tip *Looking for inspiration on subjects for a time-lapse movie? Enter "time-lapse" in YouTube (www.youtube.com) to see what other people have created.*

It might seem complicated, but creating a time-lapse movie is surprisingly simple once you know how. The long time it takes to shoot a time-lapse movie will likely go by as quickly as a time-lapse itself.

Project 16

Create Fantastic Audio for Your Movie

What You'll Need

- **Digital video camera (any)**
- **Home computer**
- **Movie editing software**
- **Cost: $100**

Audio is probably the most neglected area of home moviemaking, yet it can deliver huge payoffs with a little effort. In Project 2, you learned how to use different types of microphones to record great audio, but of course that isn't all there is to filling your movie with fantastic sound. There's more that you can do during filming and, especially, at your computer to make sure that your movie sounds as good as it looks.

In this project we'll look at camera settings for audio and also talk about techniques in post-production (meaning during editing) that you can use to improve your movie's sound, whether it involves altering the audio track to your recorded movie, adding digital sound effects, creating your own sound effects, or adding narration. You'll also learn about adding music—that great mood-creator—to add a new level of emotional depth to your movies. You can get royalty-free music online, or use programs you might already own to create your own soundtrack. Finally, we'll guide you through a project of making an enhanced audio CD, which will play video on a computer and music tracks on a stereo, perfect for when you want people to enjoy the score as much as they enjoy the movie.

See a CNET video on creating fantastic audio at http://diyvideo.cnet.com

Improving Video Camera Audio

Even if you're not using a microphone (see Project 2 for all you need on microphone tips and techniques), you can improve the sound you get from your video camera. Follow these suggestions:

- Your camera can record either 12-bit or 16-bit audio. This is a measure of how much audio data is stored during every second of filming. As you might expect, 16-bit audio sounds richer. Cameras, however, are often set to 12-bit audio by default. To check your setting, open your video camera's onscreen menu and look for the settings. Audio might be included under VCR settings (or playback settings, which is a bit confusing since you need to record the audio before you can play it back). Select audio mode and make sure it's set to 16 bit (see Figure 16-1).

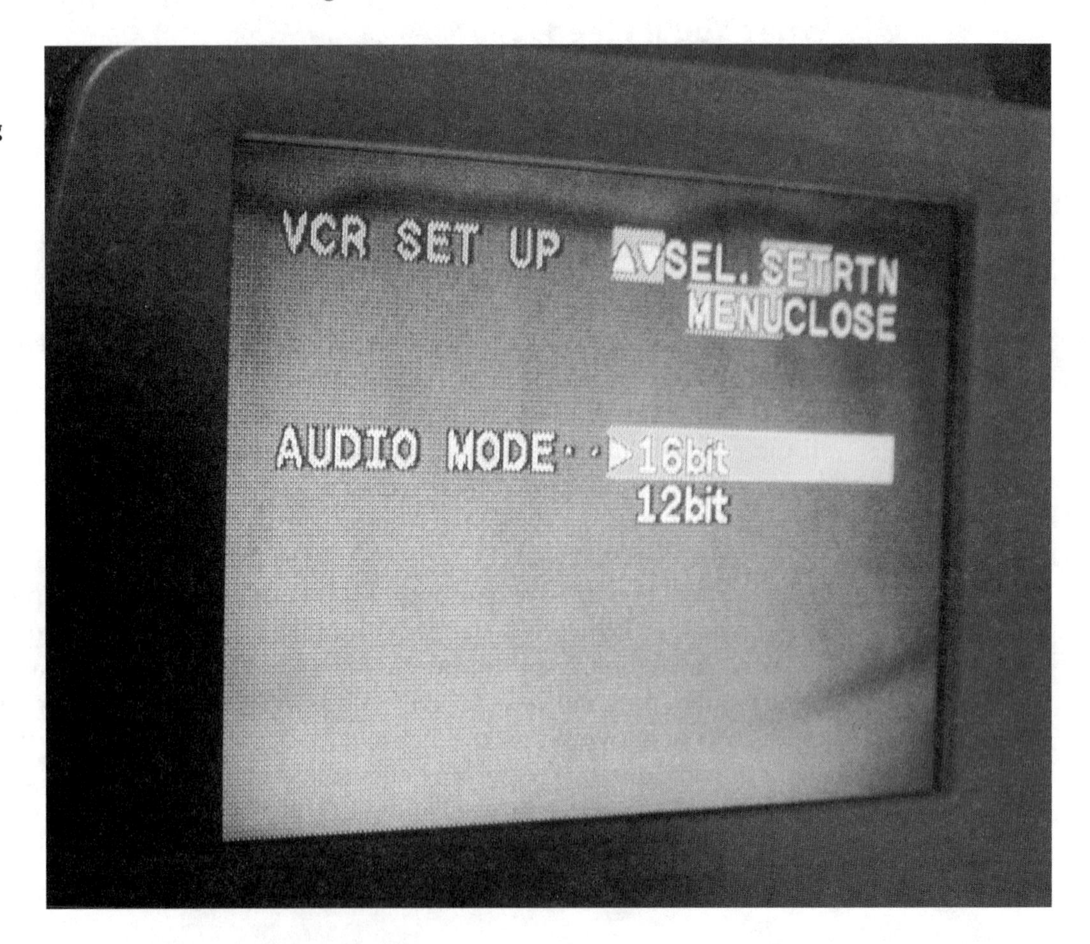

Figure 16-1
Make sure your camera is recording 16-bit audio.

- If you're the type who wants to pick up a camera and begin shooting without messing around with settings, you might not be getting the best sound every time. If you're shooting outside and there's even a moderate amount of wind, you should turn on your camera's wind screen. If you've ever had to deal with the surprisingly loud racket that wind can make when it blows against a microphone, you'll be happy to know that it's preventable. Go to your onscreen menu and look for audio controls, possibly under the VCR or playback settings. Find wind screen and turn it on when you're outdoors. Remember to turn it off when you're back inside, as the microphone is more sensitive when the wind screen is off.

note *Turning the wind screen on won't help when you're using an external microphone, just when you're using your camera's built-in mic.*

- Using headphones with your camera is the best way to monitor the sound you're recording. Your camera has a headphone jack, but it probably does double-duty as an A/V port. If you want to plug in headphones, first open the onscreen menu's audio settings and make sure the AV/Phones option is set to Phones.

Improving Audio During Editing

Your editing software also offers easy ways to improve the audio of your movie. People often get caught up trimming their scenes and adding titles, but forget that sound improvements can make watching the movie much more pleasant.

Step 1: Add Fades

If you pay attention to TV shows, you'll notice that the sound often fades in for a second at the start of a new scene. Just as editors make the visuals fade in from a black background, to bring the viewer into the scene slowly, they know that starting with full sound, especially when it's a crowd scene, is jarring. Use your editing software to fade in and out of your scenes. In Apple iMovie, when you're in timeline mode, you'll notice a purple line running through your clip on the timeline. This represents the audio level. Click that audio level one second in, to create a handle (see Figure 16-2). This is where the audio will return to its normal level. Now, click the audio level at the very beginning of the clip and drag it all the way down. You'll see that it creates a smooth curve upwards, showing how the audio will build to the full level. Do the reverse of this at the end of your clip.

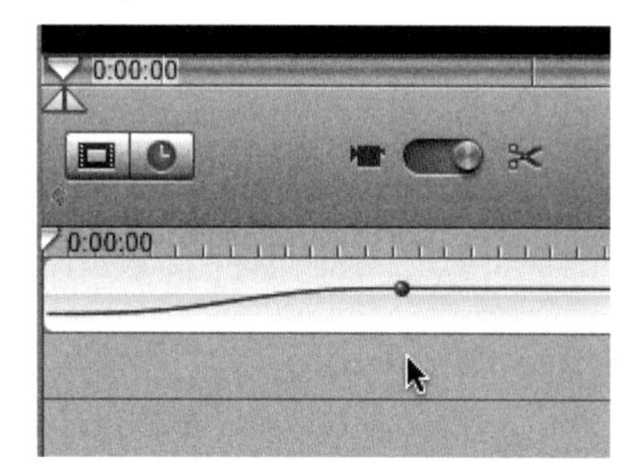

Figure 16-2
Creating a fade-in with iMovie

In Ulead VideoStudio, you need to enter the Audio View to adjust sound. Do this by clicking the speaker symbol at the lower left of the screen. This will bring up a sound menu in the upper left that lets you add fade-ins and fade-outs with one click (see Figure 16-3). You'll see the change on the audio level bar down in the timeline. You can adjust this bar by hand, if the automatic fade wasn't quite right for your scene. Note that you'll need to use the Play button in this sound settings window to preview your changes, because you won't hear them if you use the standard preview controls.

Figure 16-3 VideoStudio's audio controls

Step 2: Adjust the Audio Levels

You can use the same tools from Step 1 to adjust the audio levels for the entire clip. In Apple iMovie, you'll see a percentage listed below the timeline, next to an audio icon. Select the clip you want to change and click the icon to make the level lower or higher. Even though the starting value is listed as 100 percent, you can go up to 150 percent for an extra boost.

In VideoStudio, don't select the track on the timeline, but instead select the button in the audio controls on the left that represents the video track of your movie. Drag the audio level next to it up or down to raise or lower the audio for the entire clip. Use the Play button above the audio controls to preview your sound edit, not the Play button in the center window.

Is the sound in a scene not so hot? Perhaps a lot of harsh wind noise is ruining your scenic vacation shots? Just turn the audio way down and cover the scene with appropriate music.

Step 3: Separate the Audio

Your software will group the audio track with the video track, but you don't need to keep it that way. In iMovie, for example, select a clip in the timeline and choose

Advanced | Extract Audio. This puts the audio in a different track on the timeline. You can use this when you want to delete the audio, such as when you're making a time-lapse movie, or for when you want the audio to either continue past the video and lead into the next scene (called an L-cut) or begin before the current scene (called a J-cut).

Adding Sound Effects

When people think of sound effects, they think of crackling thunder or booming explosions—hardly material for most family videos—so they tend to neglect them when editing. But sound effects can also be a lot more subtle, and using them can add a level of realism to your work.

- Apple iMovie comes with a large number of sound effects that are good for adding ambiance, music beds, and simulated sound for many real-world objects. To open them, click the Media button on the bottom right, then the Audio button at the top right. You'll see three sound effect libraries listed first. You can drag the sounds where you want them, but for more precise placement, put the playhead exactly where you want the sound to go, click the sound, and then click the Place at Playhead button in the lower right corner of the window (see Figure 16-4).

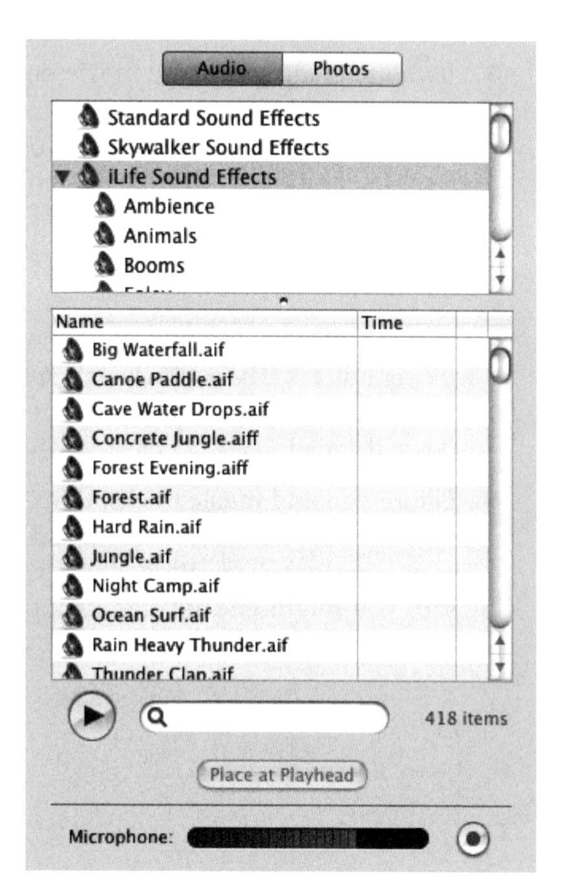

Figure 16-4 iMovie's included sound effects

- VideoStudio contains only a few sounds and they're mostly royalty-free instrumental tracks. They're also not properly labeled. To help you remember which is which, click their code names to get a text field and then enter more descriptive names.
- You don't need to stick with one sound effect at a time; you can layer them. Add a thunder sound effect on top of a rain sound effect to create a real tempest onscreen.
- Adobe Premiere Elements contains effects you can add to audio, but offers no sound effects. If you need to get some, search the Web for "free sound effect." You'll find plenty of sites. Not all of them are legit, so make sure that you have your virus software and your pop-up window blocker on before you begin.
- Can't find what you're looking for? Then make your own sound effect. Use your video camera to record it, and then separate the audio and video after you import the footage. Save the sound clip to your library, so you can reuse it in later projects.

Adding Music

There's nothing better than adding just the right song to create a mood in your video. Here are some tips for getting good results:

- If you're going to use the movie only for personal use (such as to show to family and friends), you can safely use commercial tracks from your music library. If you're posting online to a video-sharing site such as YouTube (www.youtube.com), you're probably also safe. The label or the artist could demand that the clip be removed, but so far this hasn't been happening. Getting permission is crucial, though, if you plan on commercially exhibiting the movie in any way.
- If you do plan on exhibiting your movie, use royalty-free music when creating the soundtrack. There are several sites online that sell royalty-free tracks. You'll need to purchase them upfront, but you won't have to pay any additional fees down the road.
- Before you add music tracks, think about how you want your audience to feel, since music has a large impact on mood. Adding an old French ballad over your French vacation videos is a good idea, but if the song is slow and sad, you might end up depressing your audience rather than delighting them. Choose an upbeat song and your audience will feel upbeat.

Writing Your Own Music

Can't find music that suits your video? Then create your own. Then you'll never have to worry about payments. Here are some suggestions:

- If you're using a Macintosh, your computer probably came with a free copy of GarageBand, a program that makes it simple to combine music loops and make songs (see Figure 16-5). You can even plug in a mic and sing along, or attach an electronic instrument. You'll need to export your song to iTunes and then select it in the iMovie Media section.

Figure 16-5 Creating music with GarageBand

- For truly professional results, consider investing in a more advanced program, such as Sony Cinescore (currently a reasonable $175 for download). It creates automatic music scores based on information you select.

Creating an Enhanced Audio CD

If the soundtrack to your video is really strong, you can save your movie as an enhanced audio CD, so that recipients can either watch it or listen to it: an enhanced audio CD will play in a stereo CD player just like any other audio CD, and the video will play on Windows or Macintosh computers. It won't play on DVD players and it has less storage space than a DVD, so there's a trade-off.

Although these instructions are for making an enhanced audio CD with Roxio Toast, a Mac program, you can use one of several Windows programs, such as Roxio Easy Media Creator, to make enhanced audio CDs as well.

Step 1: Save the Video

In iMovie, save your movie to your desktop by choosing Share | Share. Save it as a QuickTime movie at CD quality (see Figure 16-6). This will compress the video frame rate, although the audio won't change.

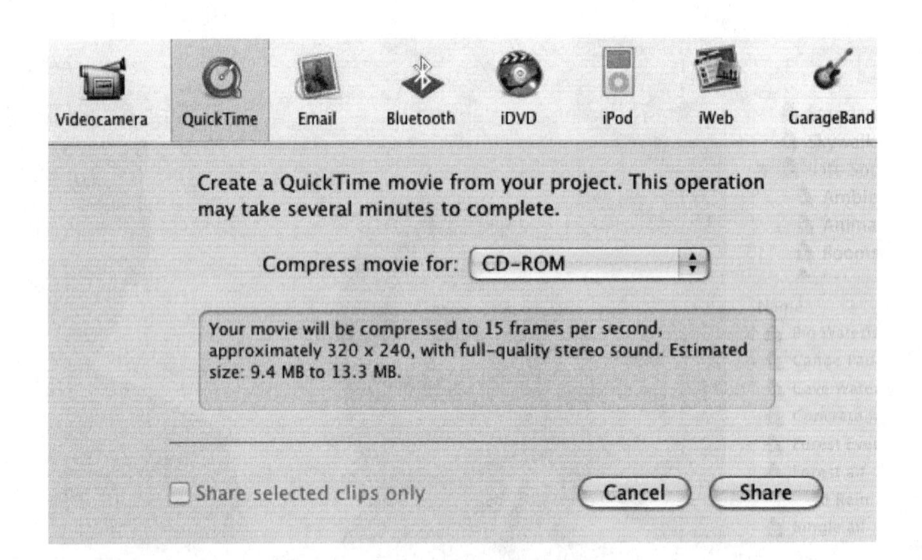

Figure 16-6 Exporting a movie in iMovie

Step 2: Make the Data Side

In Toast, select Data from the left-side menu to make a data disc, and then select Mac & PC under that, so that both operating systems can read the file (see Figure 16-7). Drag your movie file into the center area.

Step 3: Create the Music Side

From the left-side menu, select Audio and then select Enhanced Audio CD under it. Drag your song files into the center area.

Figure 16-7
Adding a video to Roxio Toast

Step 4: Go for the Burn

Insert a blank CD into your disc-burning drive and click the big red button in the lower-right corner. A CD-burning menu will open, letting you select the number of discs you'd like to create.

tip *Use CD-R discs for enhanced audio CDs, as they're compatible with more types of CD players.*

Pay as much attention to your audio as you do your video, and your audience will sit forward and pay attention as well. If you've taken the sound advice (sorry) in this project, your creations will be music to your audience's ears.

Project 17

Create a Stop-Motion Movie

What You'll Need

- **Digital camera or video camera (any)**
- **Home computer**
- **Movie editing software**
- **Tripod**
- **Cost: $200**

While digital filmmaking continues to expand, with new and more powerful tools and effects, the charm of a stop-motion movie lives on. Maybe it's because it reminds of us our childhood, or maybe because we can't help but smile when we see one. The idea of creating a stop-motion movie sounds like a lot of work, right, with days and days spent poring over a miniature cast? We've all seen behind-the-scenes specials on how time-consuming the process is, and sure, that's true if you're creating a feature film that has to look flawless on a big screen. But you can make a short stop-motion movie in your home in only an hour or so, and you'll have a blast doing it.

This is a great project to do with kids, since toys make the best stop-motion movie actors. It's also the only video project around that you can shoot with a digital camera and not lose any quality. In fact, if you have a good digital camera, it might even turn out better. We'll guide you through the process of creating this project, and show you the equipment and software you'll need.

Shooting Stop Motion with a Digital Camera

Since stop motion is all about combining static images, you can use an average digital camera to create one.

- You're going to be shooting a lot of images, so it helps to have a jumbo memory card. Consider buying a 1 or 2 gigabyte card, if you don't already have one. Even then, you might need to empty your card into your computer occasionally while shooting.
- You don't need to use a high resolution when filming these shots, as you'll likely only view them in a small video player on your computer. Put your camera on the lowest resolution so that you can fit more pictures on your memory card. (Don't use your camera's video mode; that's not the point here.)

tip *How many shots do you need for a stop-motion movie? You can go higher, but ten shots per second is a good number to start with. Your finished movie will display at 30 frames per second, so when you import your shots later, you'll set each one to play for the length of three frames.*

- Consider using human actors for your first stop-motion attempt; doing so removes the tedious process of posing inanimate objects. You'll end up with a movie that looks like a regular video, but with an interesting jerky quality.
- If you want to film someone walking, running, or performing some other fluid action, set your camera to burst mode and use the slowest interval. This lets you take several shots in quick succession.

tip *By using humans to make a stop-motion movie, you can pull off some interesting effects, such as having someone jump and stay in the air an extra-long time.*

Importing Still Photos

Once you've shot the hundreds (or thousands) of frames you'll need, it's time to import them into your video editing program. The following steps are for Apple iPhoto and iMovie, but you can use other applications in much the same way.

Step 1: Create an Album

Connect your digital camera to your computer and import all of your photos into iPhoto. Create a new album by clicking the "+" sign in the lower left and give it a name. Drag all of the stop-motion pictures you took into the new album.

Step 2: Linking Up

Open iMovie and click the Media button in the lower right, then click Photos in the upper right. Scroll down to see the album you just created.

Step 3: Grab 'em All

Click the album name and you'll see all of your photos displayed just below the list of albums (see Figure 17-1). Click the first photo, hold the SHIFT button down, and then scroll down and click the last photo. By doing this, you can adjust the properties for all the photos at once.

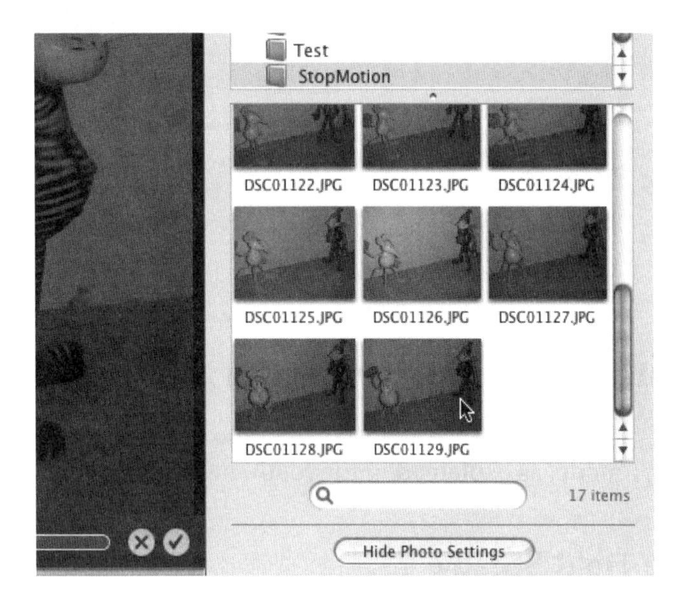

Figure 17-1

Import your still shots into iMovie.

Step 4: Set the Options

When you click the first photo, a transparent pop-up window displays, giving you photo options, as shown in Figure 17-2. If it doesn't open, click the Show Photo Settings button just below the photos. Start by turning off the Ken Burns effect by unchecking the check box. You don't want photos panning and zooming during a stop-motion movie. Set the top slider, which shows the image size, all the way to the left. This ensures that the whole image shows up. Drag the lower slider all the way to the left, as well, so that each photo goes by as quickly as possible. The speed will read 0:03. This might seem odd if you're not used to video timecode standards, but the first part, the number before the colon, is a measure of seconds, and the number after the colon is a measure of frames. Remember that there are 30 frames to each second of video (or 29.97 frames, if you need to be exact). Since the number after the colon is 03, each image will display for three frames, or one-tenth of a second, and ten shots will make a whole second. This is as fast as you can display your images with iMovie.

Step 5: Move to the Timeline

Click Update, and the pictures you highlighted will automatically be added to your timeline. You can now preview, add music or narration, or add titles as you would with any other movie.

Figure 17-2
Setting iMovie's photo options

Step 6: Save It

To share your movie on your desktop, choose Share | QuickTime and choose to save your movie at full quality.

Shooting Stop Motion with a Video Camera

You can also use a video camera to take your shots, but you'll be using the still photo mode, as explained in the following steps.

Step 1: Hold Steady

Most of the time when you're shooting stop-motion movies, you'll want to secure the camera with a tripod first. Having your camera move by even a fraction of an inch can ruin a scene and all the work that came before it.

If you're using toys as your actors, consider using a toy car or skateboard as your dolly. Mount the camera on it and then move the camera forward between shots to create a smooth dolly shot.

Step 2: Use Your Remote Control

Your video camera probably came with a remote control that you've never used, right? Well dig it out, because now is the time to use it. A remote (such as the one shown in Figure 17-3) lets you take shots without touching the camera, which is a big help when you're trying to keep the camera still.

Step 3: Set Your Shots

Pose your subjects and make minute movements between shots. Remember that each shot will be on the screen for only three-tenths of a second, so the action in each shot should be fairly small.

tip

Apple iMovie doesn't have stop-motion capabilities, so Apple users who want advanced features such as onion skinning (which allows you to view transparent versions of previous frames when shooting a new frame) should look to iStopMotion (www.istopmotion.com), a great download for only $39.95.

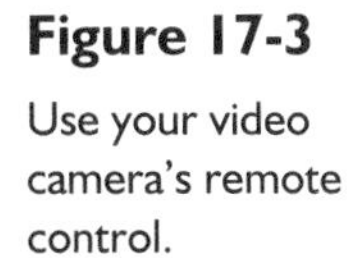

Figure 17-3

Use your video camera's remote control.

Creating a Stop-Motion Movie with Adobe Premiere Elements

When Adobe Premiere Elements 3 came out, it included the fun surprise of being able to create stop-motion movies, making it the first and only of the consumer-level video editing applications that can do so. Sorry, but if you have version 1 or 2, you're going to need to upgrade to create stop-motion videos.

tip *Shoot your movie in an area with even lighting. Completing even a short movie will take an hour, and you don't want the light fading as you're working.*

Step 1: Make a Connection

Start by connecting your video camera to your computer, with the cable that came with your camera, and put your camera in record mode. Open Premiere Elements.

tip *Connect your camera's power cord when shooting stop-motion videos. Your battery probably won't last the whole time.*

Step 2: Get Media

Click the Get Media From button in the upper-left corner and choose the DV camcorder option. A large Capture window will open.

Step 3: Time for the Timeline

Click the More button in the upper-right corner and make sure Capture to Timeline is checked, as shown in Figure 17-4. This will put your shots in the application's timeline automatically.

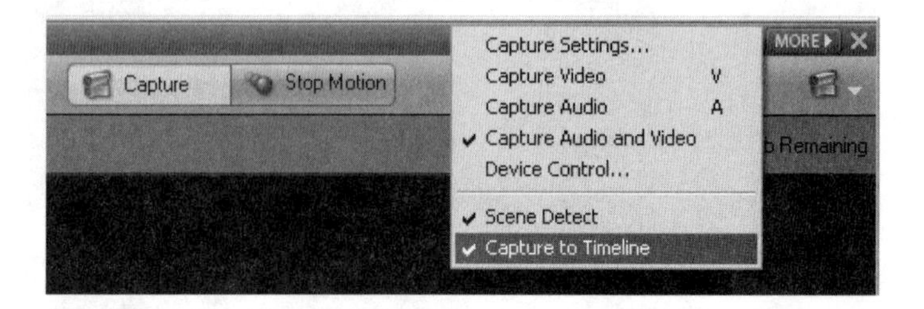

Figure 17-4

Options under the More button

tip *The Joby Gorillapod (www.joby.com), also mentioned in Project 4, is a great tripod to use when shooting a stop-motion movie. It's small and flexible, letting you get in tight and hold the video camera at awkward angles.*

Step 4: Stop Motion Capture Setting

Click the More button once again and select Device Control. A Preferences window will open, as shown in Figure 17-5. Select Stop Motion Capture on the left side.

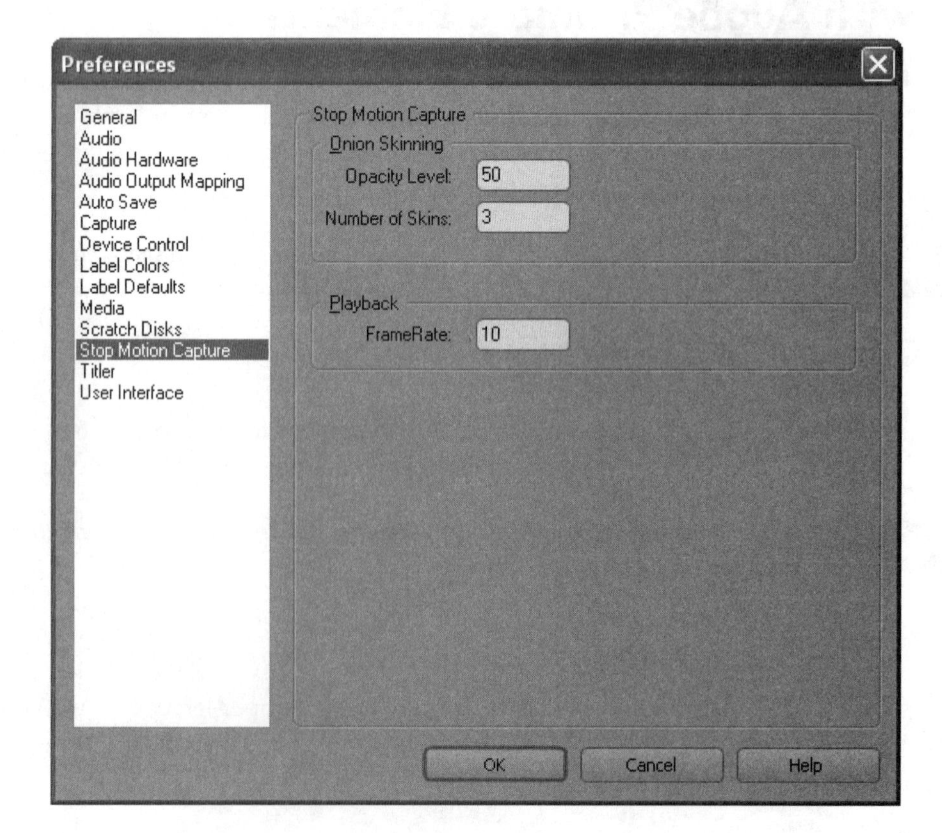

Figure 17-5

The Stop Motion Capture setting.

Step 5: Skinning the Onion

In the Preferences window, you can set the number of onion skins that you see when shooting. Onion skins are transparent versions of frames you've already shot. Having them superimposed over your image lets you see how the characters have already moved. Five is the default value, although some people prefer three to keep the image less crowded. This window also lets you set the opacity of your onion skins and the number of frames that will display during every second of the video. The default value is 12, but set it to 10 to save yourself a little work. Click OK when you're done.

Step 6: New Movie

Click the Stop Motion button at the top of the Capture window. Choose the option to create a new stop-motion movie.

Step 7: The First Frame

You'll see your camera's image on your computer screen (see Figure 17-6). Pose your figures and click the Grab Frame button when you're ready. Your image will be saved to the Available Media panel of the main interface.

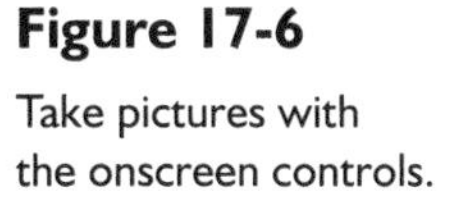

Figure 17-6
Take pictures with the onscreen controls.

Step 8: Do It Again

Re-pose your figures. You'll notice that a translucent version of your first shot stays on the screen. This is the onion skinning feature, which helps you line up successive shots. When you're ready, click Grab Frame again.

If you're shooting a stop-motion movie with a lot of characters in it, be sure to re-pose all of them each time, so that they're all in motion. You don't want one character awkwardly standing still while everything else in the scene is moving.

Step 9: ...and Again

Continue to pose your figures and take new shots. It will take ten shots to make one second of video, so you have a lot to shoot.

Step 10: Preview Anytime

To be sure you're getting the results that you want, check the Preview check box in the lower-right corner of the Capture window, and then click the Play button in the bottom center. When you're ready to shoot more frames, uncheck the Preview check box.

Don't like what you just shot? You can delete the last frame at any time by clicking the trash can icon in the Capture window's lower-left corner.

Step 11: Make a Movie

When you're finished shooting, click the Close button (the X) in the top-right corner. You'll be asked if you want to save your frames as a movie file. Click Yes and save it to your hard drive. Your movie is now saved and also added to your timeline.

Continuing at a Later Time

Creating a stop-motion movie can be time consuming, and you may not be able to shoot your entire movie at one time. Luckily, Premiere Elements has you covered.

Step 1: Get Media

On your return session, again click the Get Media button in the top-left corner and choose the DV camera option. The Capture panel will open.

Step 2: Click Stop Motion

Click the Stop Motion button in the top center.

Step 3: The Last Frame

Your Available Media panel should be visible to the left of the Capture window, as shown in Figure 17-7. Grab the last frame you shot in your previous session and drag it into the Capture window. You'll see an onion skin image of the previous frame overlaid on the current view. This will help you line up your current shot.

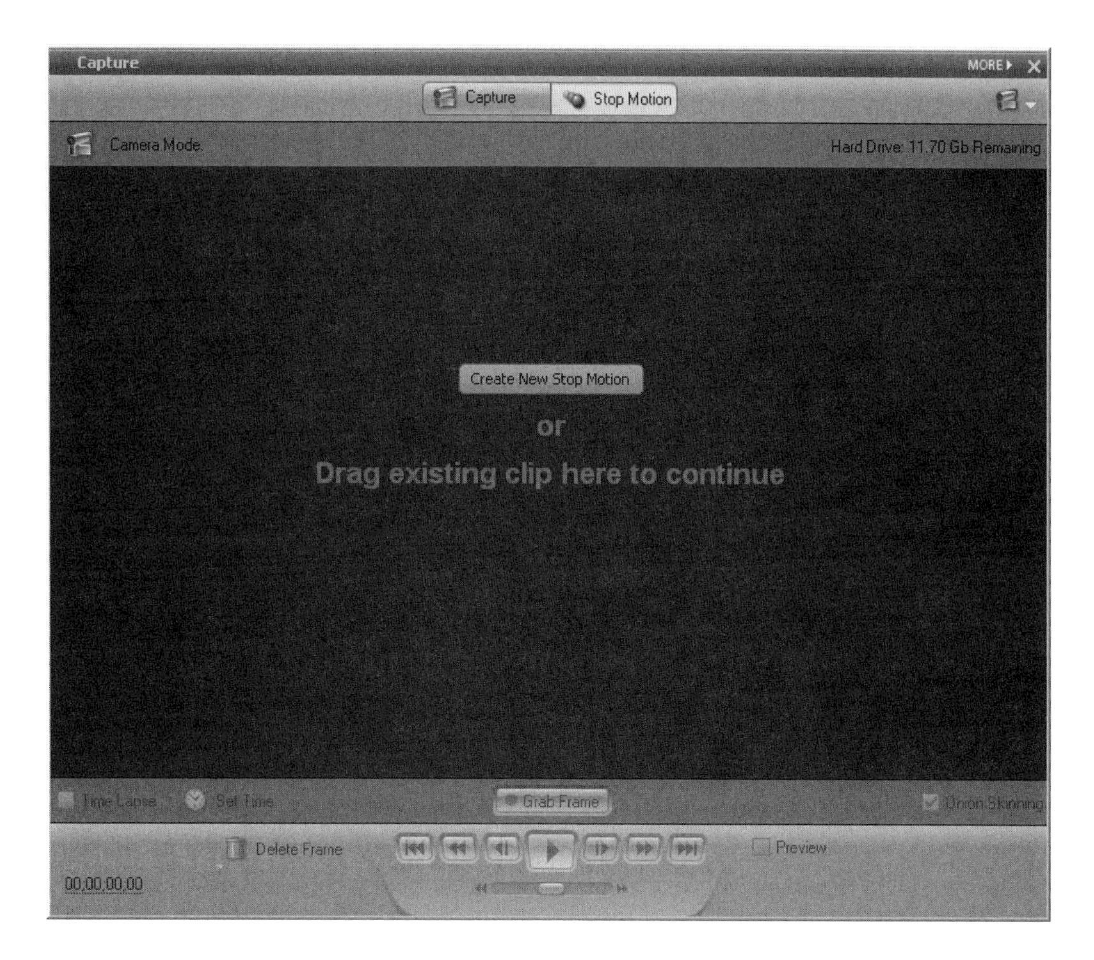

Figure 17-7 Grab the last frame from your previous session.

Step 4: New Shots, Old Numbering

Click the Grab Frame button in the bottom center when you're ready. Your new shots will be numbered correctly, so that they automatically follow from the previous shots.

note *LEGOs make an ideal cast and set for stop-motion movies. If you agree, visit Brickfilms (www.brickfilms.com), a large online community dedicated to making and sharing LEGO movies.*

When you're done, no matter how simple or jerky your stop-motion movie comes out, your guests will be begging to watch it. As mentioned earlier, there's just something about stop motion that makes people smile.

Project 18

Create a Video Holiday Letter

What You'll Need

- **Digital camera or video camera (any)**
- **Home computer**
- **Movie editing software**
- **Tripod**
- **Cost: $200**

This project marks a break in this book. It's the last of the "creative" projects, which emphasize a range of shooting and editing techniques, before the projects get more mechanical, focusing on things you can do at your computer or with your home entertainment system. As such, we're making it an all-star jam: a delicious holiday stew full of many of the tricks and techniques we've already covered. Think of it as a graduation project in creative home video, a chance to pull together many of the things you've learned to create a DVD that will really entertain your viewers.

The project this time is to create a video holiday letter. Many people send a letter out with their holiday cards, letting family and friends know what they've been up to for the past year. With a video holiday letter, though, you can not only tell them but show them as well. This project will guide you through five fun projects that will involve the whole family, show off a few striking video shooting and editing techniques, and delight your recipients a lot more than another holiday letter will. Although we'll be using Adobe Premiere Elements for all of the instructions, you can create similar effects with other editing programs.

The Catch-Up Video

The primary purpose of a holiday letter is to let your friends and family know what you've been doing all year, so that's the idea behind our first segment. But here you won't just tell what you've been doing, you'll show it.

Step 1: Film the Letter

Plan out what you're going to say before you begin, so you're not left scrambling and ad-libbing on camera. You want to tell your viewers what the family has been up to. You're also going to show them. Since you probably haven't been carrying a video camera with you all year, you'll create humorous "dramatizations" to show your highlights.

Start with the whole family sitting together on a couch or at the dining room table. Make sure everyone fits in the shot nicely. Set up your video camera and rotate its viewing screen 180 degrees so that you can see the shot from in front of the camera. Have each person speak in turn, telling what they've been up to that year. Each person's speech should end with them telling what their highlight was, then letting their voice trail away as they remember it. You'll add the dramatizations during editing at these moments.

Step 2: Flash Back

For the dramatizations, or flashbacks, get a little silly. Use costumes, wigs, and fake moustaches. For example, if sister's highlight was making the dean's list at college, film her furiously studying and writing in her notebook. Then have the "dean" walk into the scene (perhaps Dad in a fake moustache) and announce to her that due to her excellent effort, she's made the dean's list that semester. Have everyone rush into the scene and begin celebrating, throwing confetti and blowing noisemakers.

Step 3: Make it Black and White

Once you've got all your segments, import them into Adobe Premiere Elements. We're going to use the program's title themes to tie them all together and add to the comedy. Break the video into scenes, if necessary, and open the Effects and Transitions menu from the Media panel in the upper-left corner. Open the Video Effects section and then open Image Control. Find the Black & White filter (see Figure 18-1). Drag this filter onto your dramatization clips to make them black and white, creating the illusion of remembering back in time.

Step 4: Frame It

Open the Title Templates menu from the Media panel and open the Holidays and Events folder. Open the Blinking Lights folder (see Figure 18-2). You may have to wait a few seconds for the options to show up onscreen. Grab the frame option and drag it to the video track just above your movie. Don't put it on the same video track, or else it will come before the movie instead of putting a frame around the first scene. Click the movie title in the preview window to change it, and add your own title.

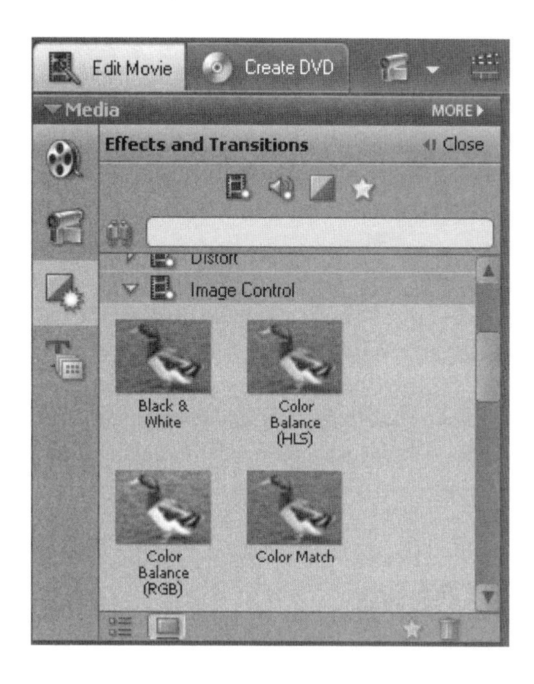

Figure 18-1
Premiere Elements' Black & White filter

Step 5: Add a Subtitle

Insert your playhead at the start of the first black-and-white flashback clip. Grab the lower-third blinking lights title option and put it in the video track just above your first flashback scene. This will give you a matching title along the bottom of the screen. Drag the title's right edge, to make it as long as your flashback. Click the title in the preview window and type **Dramatization**.

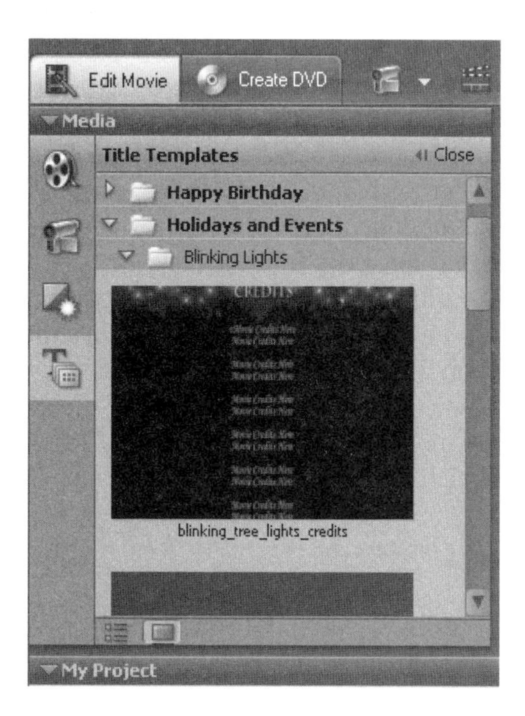

Figure 18-2
Premiere Elements' Holidays and Events templates

Step 6: Freeze It

Position your playhead at the last frame of the movie and create a freeze frame by clicking the Camera button just above the preview window. Indicate how long the freeze frame should last. You'll then see the freeze frame clip in your timeline. Return to the titles section and click-and-drag the blinking lights credits option to above your freeze frame. Stretch it to last the whole length of the frame. Add the right names to your credits, and then go to the Get Media section of your Media panel and import a holiday song. Drag it into the audio track to provide a little music over your title and end credits.

tip *Do you find video playback to be slow and choppy with Premiere Elements? Just shrink the video preview window by grabbing its right edge and pulling it to the left and playback will be much smoother.*

A '50s Sing-Along

For the next segment, we're going to mimic the look of an old '50s Christmas TV special and have the family sing together. We'll use the narration tool to add some comedy.

Step 1: Set the Scene

Have the family dress in holiday sweaters and create an old-time scene in your living room, perhaps with Dad reading the paper and smoking a pipe and Mom in a holiday apron. Have your family use stilted dialogue to say what a great Christmas it's been, and then lead the family in singing their favorite carol. It doesn't matter how the singing sounds, because you're going to remove the audio from this part.

Step 2: Tint the Color

Import the footage into Premiere Elements and drag your video to the timeline. Open the Effects and Transitions menu in the Media panel and click to open the Image Control options. Scroll down until you see Tint, as shown in Figure 18-3. Drag the effect onto your clip in the timeline.

Look for the Tint entry in the Properties menu, in the upper-right corner. Click the triangle next to its name to open it. Click the white box next to the Map White To option. You'll get a color picker on the screen. Select a faded yellow color and then click OK. Change the value for Amount to Tint, just below the mapping option, until you like the look in the preview window. This will give your video a yellow cast, making it look faded and old.

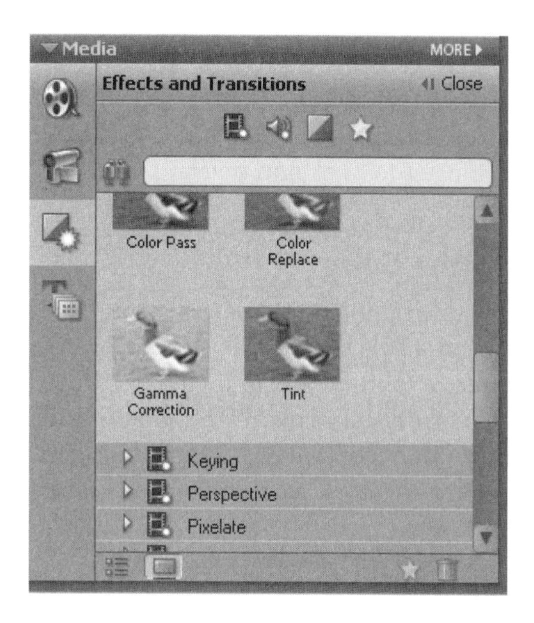

Figure 18-3
Select Tint from the Effects and Transitions menu.

Step 3: Make Some Noise

Go back to the Effects and Transitions menu and scroll down until you see Stylize. Open it and look for the Noise filter. Drag it onto your clip. In the upper-right Properties panel, open the Noise properties (see Figure 18-4). Adjust the Amount of Noise value until you like what you see on the screen. This will help create the look of old film.

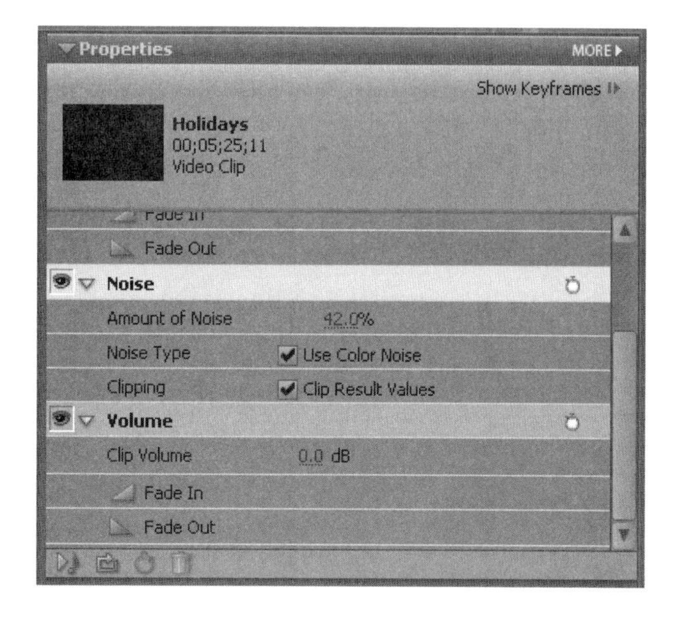

Figure 18-4
Adjust the Noise properties.

Step 4: Scratch the Sound

Put your playhead just at the point where the family starts singing, and then split the clip in two. Right-click the audio track and choose Unlink Audio and Video. Then, click the audio track and delete it.

Step 5: Redo the Song

Plug a microphone into your computer and have the family gather around. Place the playhead at the start of the singing scene, and then click the Add Narration button just above the timeline and click record. Have the family all sing their parts again, at the correct moments. Don't worry if the lip sync is off; you want it to be to add some humor. Stop the recording when you're done. Your singing will be added to the timeline's narration track; scroll all the way down to see it.

Sledding Action

Enough with all of the indoor shots; it's time to get out in the frosty air and shoot some sledding. We'll use time effects to make the footage fun.

Step 1: Take a Ride

Take the whole family to a sledding hill or a snow tubing center. Take a lot of footage of everyone going down the hill (and that includes the person who usually holds the camera). Also, carry the camera as you sled down the hill, for an action shot. Try to get a few dramatic moments, like someone jumping over a ramp or a collision on the hill.

Step 2: Action!

Import the footage into Premiere Elements and edit the footage into a lot of quick cuts showing family members zooming down the hill. Arrange the clips to suggest a day of nonstop action. Keep the whole video short, around two minutes in length.

Step 3: Racing Titles

Open the Title Templates section of the Media panel, open the Sports menu, and then open the Racing section. You'll see titles designed for car-racing movies. Drag one of the titles to the beginning of your clip and place it in a separate video track. Click the text in the preview window to change it.

Step 4: Go Slo-Mo

Use the Time Stretch tool to slow down especially dramatic moments, like a wipeout or a jump of a snow ramp (see Figure 18-5). To do so, move the clip to the right of the one you want to stretch further right on the timeline to make a little room. Click the

Time Stretch tool just above and to the left of the timeline. Drag the right edge of the clip you want to slow down to the right. This will help highlight the action and put some variation into your clips.

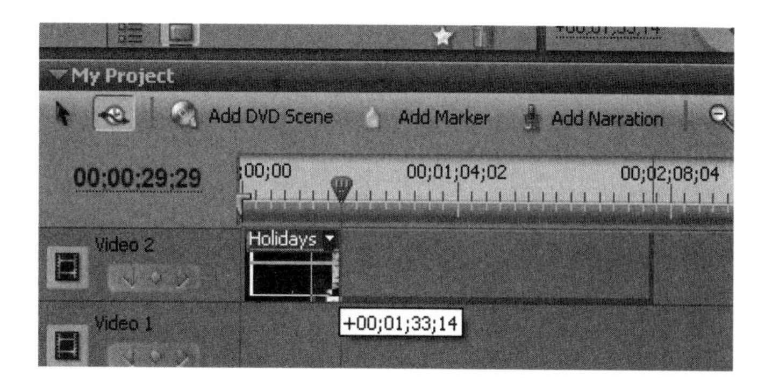

Figure 18-5
Use the Time Stretch tool to slow down the action.

Step 5: Add Music

Open the Get Media section of the Media panel and import some songs appropriate for a day of sledding action. Perhaps the James Bond theme song, or "Dead Man's Curve," by Jan and Dean, might be fitting. Drag the songs into the audio track. Don't use the whole songs, but just sections of each. This will create a sense of movement, as one song blends into another.

Vacation Highlights

Telling about your vacations is a big part of holiday letters, so we'll give vacation highlights their own segment. You'll use chroma key effects to dramatically stand in front of your images.

Step 1: Chroma Key

Set up a solid-colored panel, preferably in bright light blue or green, that you can speak in front of. This doesn't need to be a full-body shot, so the panel doesn't need to be that large; it just has to fill the frame. For much more information on creating a chroma key background, see Project 11.

Step 2: Compose the Shot

Have someone in the family stand in front of the panel and film them as they tell about your vacation highlights from the year. Don't put the person at the center of the shot, but rather have them stand about one-third of the way in on the left side. This will create room for your pictures.

Step 3: Remove the Background

Import the footage into your software and drag it into the Video 2 track of your timeline. Open the Effects and Transitions menu from the Media panel. Scroll down until you see Keying, and then click to open it. Click-and-drag the Chroma Key filter onto your clip. Go to the Properties panel in the upper-right corner and open the Chroma Key properties (see Figure 18-6). Use the eyedropper tool to select the color you'll be keying out.

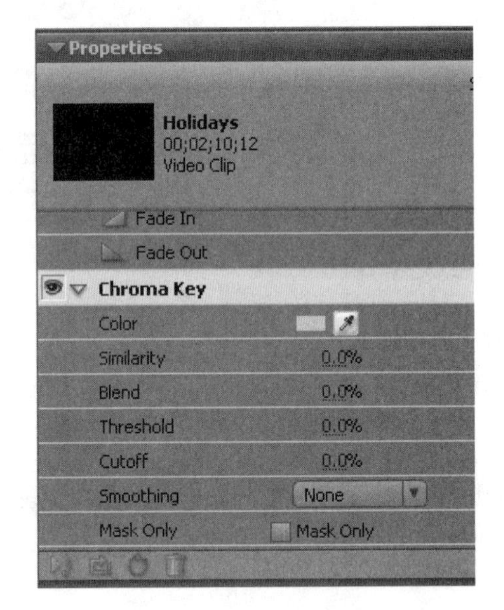

Figure 18-6

Select the background color with the eyedropper tool.

Step 4: Add Images

Go to the Get Media section of the Media panel and import your vacation photos. Drag them into the Video 1 track of the timeline. They'll automatically show up behind your speaker. Arrange the photo clips in the timeline so that they correspond with what your speaker is saying; you can even drag a photo's right edge to make it display longer, if necessary.

The Time-Lapse Christmas Tree

It's going to begin to look a lot like Christmas in a hurry with this last segment. Set up your video camera near where you'll be putting up your Christmas tree, and attach it to your computer. If you have a digital camera with interval shooting, you can use that without a computer connection. Otherwise, you'll need to connect your video camera and PC to use Premiere Elements' time-lapse effects. Be sure to go over Project 15 first to understand time-lapse shooting.

Step 1: Tripod Shot

Set up your camera with a tripod (an essential, since the camera can't move a bit during shooting) to show the whole area where you'll be setting up your tree. Connect your camera to your PC and put the camera in record mode.

Step 2: Grab from the Camera

Click the Get Media From option in your Media window and select the video camera option. A Capture window will open. Click the More button in the upper-right corner and make sure Capture to Timeline is checked.

Step 3: New Movie

Select Stop Motion from the top middle of the window and choose the option for a new Stop Motion movie.

Step 4: Interval Shooting

Click the time-lapse button in the lower-left corner. You'll be able to set the time interval for grabbing new frames (see Figure 18-7). Since setting up your decorations will move quickly, choose a quick interval, such as one frame ever two seconds.

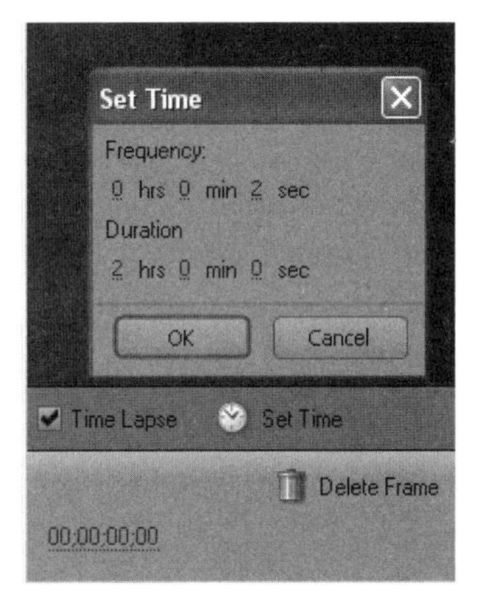

Figure 18-7
Set the interval for time-lapse shooting.

Step 5: Make a Movie

Click the Start Time Lapse button in the bottom center. When you're finished, you can simply close the Capture window. You'll be prompted to save the images as a movie file.

When you're all done, you'll have five fun segments that will delight your recipients. All you need to do is create a DVD menu and burn the segments to discs. For much more on creating attractive DVD menus, see the next project.

Project 19

Create Hollywood-Style DVD Menus

What You'll Need

- **Home computer**
- **Movie editing software**
- **Cost: $100**

Creating a great-looking DVD menu is the cherry on top of the sundae. You've already spent hours filming and editing a movie, and now you get to design a DVD menu that puts a professional finish on your work. A great DVD menu will have your viewers saying "Wow!" before they even see the first frame of your video. That's because today's editing and disc-burning programs allow you to create menus with full motion backgrounds, motion buttons, background music, and scene menus. You can even designate a short movie to play automatically when the disc is first inserted, before the DVD menus show up, if you like.

In this project, you'll learn DVD menu creation for three programs: Apple iDVD, Adobe Premiere Elements, and Ulead VideoStudio. Each offers its own selection of themes and templates, letting you create menus that match the subject of your movies. Put a little time into getting to know the features of whatever program you're using, because nowadays nearly every DVD-creation program contains slick features that will make your finished disc look just as professional as the commercial DVDs you buy in stores. Invest time in creating a Hollywood-style menu and you might even surprise yourself.

Creating Menus with Apple iDVD

While some disc-creation programs are bundled together with video editing programs, Apple chose to divide the two into separate applications. iMovie can hand a finished video off to iDVD, so iMovie is the place to start creating your DVD menu.

Step 1: Add Markers

When you're finished editing your movie in iMovie, it's time to add chapter markers. These will create scene markers when you import the finished video into iDVD. Put the playhead at the very start of your movie in the timeline. Click the Chapters button in the lower-right corner of the window, and then click Add Marker just above it. You'll see a marker added in the column above, with an image showing the scene (see Figure 19-1). The marker will take the name of your project by default, but you should give it a unique name, because the scene markers in the DVD won't be much good if they all have the same name.

Figure 19-1

Add chapter markers in iMovie.

tip *Having trouble putting the playhead in exactly the right place? Just get it near the correct frame, and then use the left and right arrows on your keyboard to move it one frame at a time.*

Step 2: Mark the Major Scenes

Move the playhead to the start of the second scene of the movie. You're creating a menu that will let people jump to exactly the scene in the DVD that they want to view.

Did you skip past a scene that you now think you should include? Don't worry—you can add a chapter marker as you normally would, and iMovie will put it in the correct place in the list.

You don't need to create a marker for every scene; consider where they will be the most useful to viewers. Click the Add Marker button again and name your marker. Repeat this process for all the scenes you want to add.

When you create a chapter marker in iMovie, you have the option of adding a web link (see Figure 19-2). This is useful when you're going to put a QuickTime movie online or create a video podcast. The web link will show up on the bottom of the screen for eight seconds, with the chapter name used as the link name. Viewers can click it to go to the set web page. Web links don't show up in DVD projects.

Figure 19-2

You can add web links to your videos with iMovie.

Step 3: Switch to iDVD

When you've finished adding markers, choose Share | iDVD. Your video will be loaded into iDVD automatically.

Step 4: Choose from a Menu of Menus

Click the pop-up menu in iDVD's top-right corner and select All to see all of iDVD's menu options. Many of iDVD's menus are grouped into themes. This lets you create matching top menus and submenus. If the menu has a gray triangle next to it in the menu list on the right, it has a matching submenu. Click the triangle to see the variations. If the menu preview window shows a round icon of a person walking, it means that menu contains motion, which will play when the menu is open. Using motion menus is an easy way to add a professional touch to your DVD. Scroll through the DVD menus and click the ones you'd like to see at a larger size. Find one you like and select it by clicking it.

note *With iDVD 6, Apple introduced widescreen DVD menus. You'll see them at the top of the menu list. If you import a standard-resolution movie into a widescreen menu, the program will give you the option of making your movie widescreen, as well. Doing so will result in the top and bottom of your video being clipped away, so I don't recommend doing it.*

Step 5: Use the Map

Click the button in the bottom center of iDVD that looks like a flowchart (see Figure 19-3). This will show you a map of the entire structure of your DVD, including subpages. To add a menu template to a subpage, click the page and then click the template you'd like from the list on the right. To see a large version of your finished subpage, highlight it in the menu and then click the map button again. You can mix in different templates, but it looks better to stay with one look throughout.

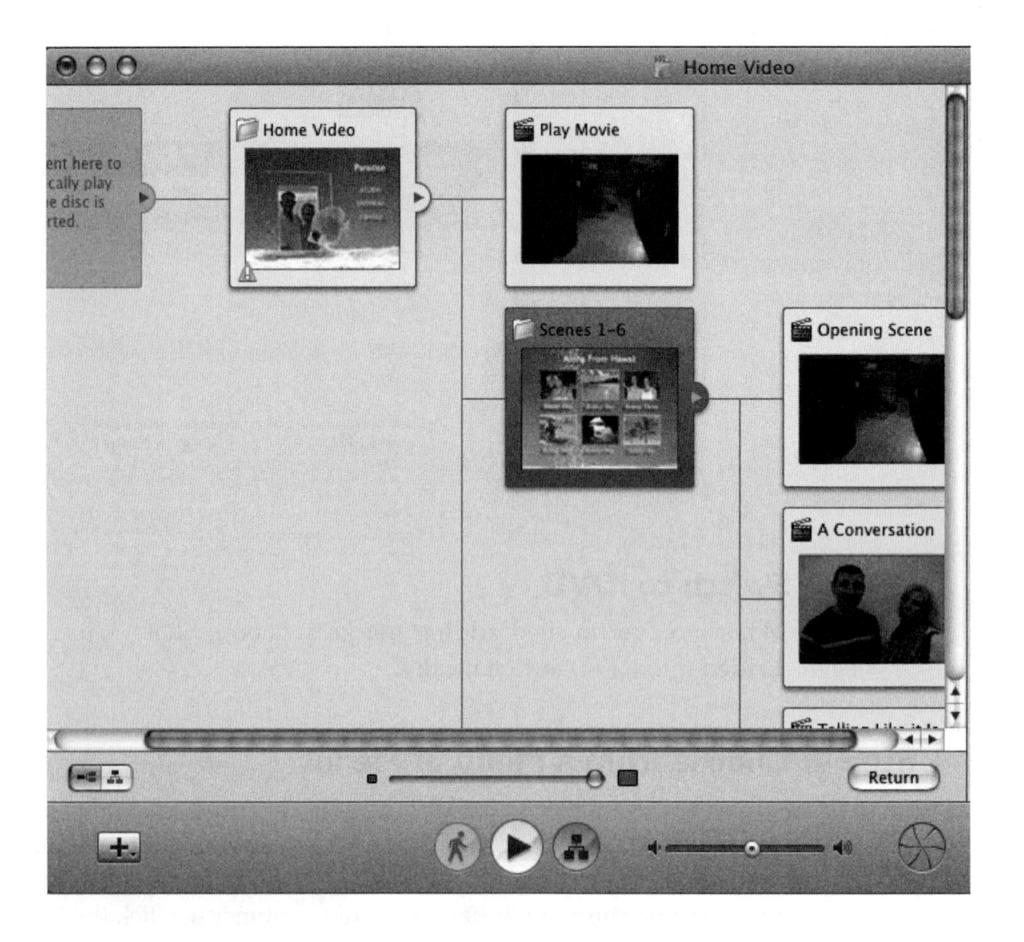

Figure 19-3

Click the map button to see a visual display of your DVD menus.

note *The first box in the map is for auto-play content. Drag a short video here so that it plays before the menu.*

Step 6: Fill In the Drop Zones

The various templates have drop zones where you can add your own movies or photos to personalize the look. You'll see in the preview window where you should add media. To add your own movies or photos, click Media in the lower-right corner to get a list of the content stored on your computer. Select either Photos or Movies at the top of the window to find the images or videos you want to use, and then click-and-drag them into the drop zones. You can also view your audio tracks in the Media section to choose a background song that will play when that menu is displayed. To choose a song, simply drag it into the preview window.

Step 7: Fine-Tuning

Click the Menu button in the lower right to personalize the menu you have open on the screen (see Figure 19-4). If you've added a background video, you can set how long it plays before it loops back to the beginning, and you can set the volume of the background music track. You can even change the font and color of your onscreen text. Select the text you want to change by clicking it in the preview window, and then use the customization controls to customize it.

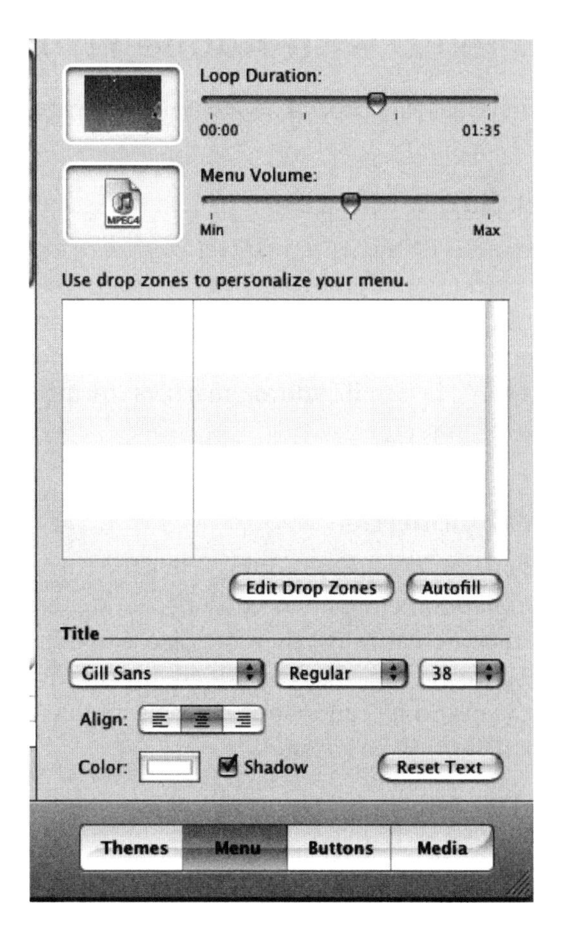

Figure 19-4 iDVD's menu personalization options

Step 8: Test It Out

Be sure to preview your menus when you're finished, so that you know exactly how they'll look on a TV screen. To do so, click the Preview button in the bottom center. You'll get a virtual DVD remote, which you can use to move around the menus just as you would with a real remote.

You can change the look or size of any button in your DVD menu by clicking the Buttons button in the lower right. To add a hidden video to your disc, remove the button's picture and make the text label simply a period. Then make the font color match the background.

Step 9: Go for the Burn

Finally, put a blank DVD into your disc drive and click the Burn button in the bottom center of the iDVD window. If you've missed any drop zones, iDVD will helpfully warn you before you create a disc.

If you're burning a DVD with a notebook computer, be sure to plug in the power cord first. Burning can take a long time, depending on the size of your movie, and you don't want your notebook running out of power in the middle.

Creating Menus with Adobe Premiere Elements

Here's how the process works with Premiere Elements:

Step 1: Set Some Markers

Premiere Elements' advanced controls (see Figure 19-5) let you set different types of markers within your video, so that you can automatically split one video into multiple movies when you create a DVD from it, and create scene menus as well. To read much more about using different types of markers, see Project 8, which covers the topic in detail. Adding markers during the editing process makes DVD creation much simpler.

Step 2: DVD Controls

When you're finished editing, click the Create DVD tab in the upper-left corner. If you don't add a template to your disc, you'll create an auto-play DVD, where the video starts automatically when the disc is inserted into a DVD player. Any main markers or scene markers that you added during editing will let viewers use the Next Scene or Previous Scene commands on their remotes to jog through the movie. Stop markers, however, will be ignored.

Step 3: Choose a Template

When you click the Create DVD tab, the list of DVD menu templates will display on the left. Every template in the program has both a main menu and a submenu, which

Figure 19-5
Add markers to your finished movie.

is great for creating a unified appearance. If the template name has an "(A)" in front of it, you can add your own audio track; if it has a "(V)" in front of it, you can add your own video for the background. Drag the template you want into the center DVD Layout panel.

Step 4: Button Control

Change the arrangement of your menu buttons by dragging and resizing them in the center DVD Layout window. You'll see more settings on the right, in the Properties menu (see Figure 19-6). Here, you can specify whether or not you want your buttons to show motion, and where in the video your button's motion should start.

Step 5: Customize the Look

If you've chosen a template with customizable audio or video, you can drag in songs, images, or videos. You get these files from the Media panel on the left. If you've already loaded the files you want into your project, click the Available Media icon to show them. If you need to add media, click the Get Media From icon and load the files that you want to use. Then, drag them into the DVD Layout window.

caution *DVD+R and DVD-R discs are compatible with more types of DVD players. Choose them when you're shopping for blank discs.*

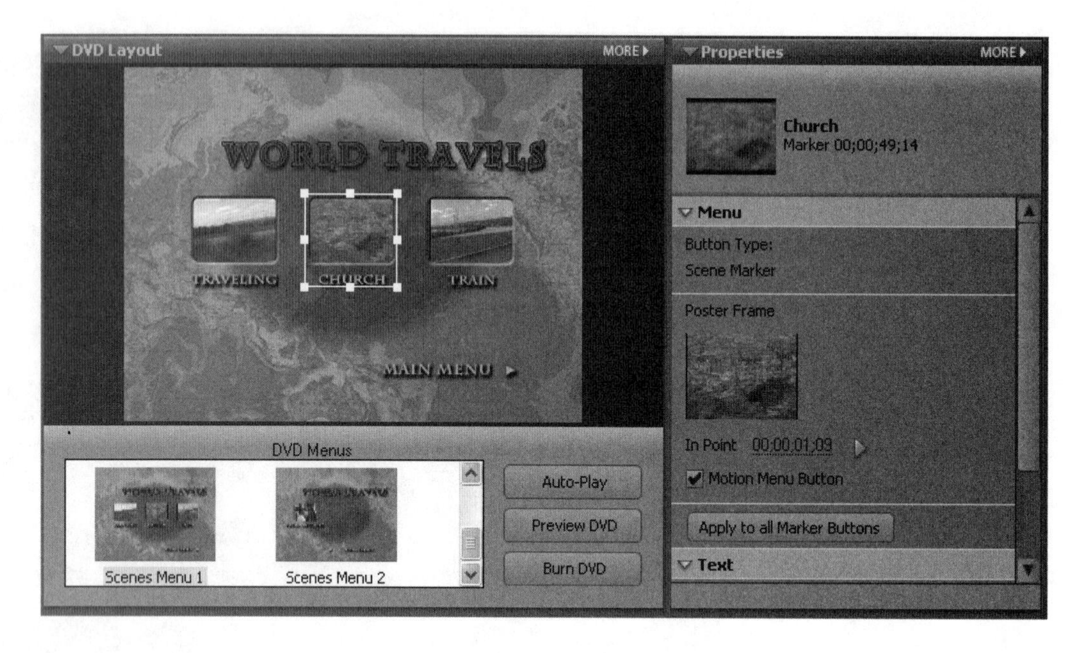

Figure 19-6
Button formatting options in Premiere Elements

Step 6: Burn Your DVD

When you're finished with all your settings, load in a blank DVD and click the Burn DVD button in the center of the screen.

tip *Will you be sending your finished disc to a person in another country? If so, you should know what type of DVD format you need to use for them. North America uses the NTSC format, as do Central America, part of South America, Japan, and the Philippines. If you're sending your disc to an NTSC country, you don't need to do anything differently. Europe, the Middle East, Australia, and much of Africa, Asia, and South America use the PAL format. If you're sending your disc to a PAL country, Premiere Elements lets you change the format to PAL in the disc-burning options that come up after you click the Burn DVD button.*

Creating Menus with Ulead VideoStudio

Many of the lower-priced video editing suites don't have an attractive library of DVD menu templates, so it's something to look into before you buy. VideoStudio's menu templates are well designed and allow for motion buttons.

Step 1: Chapter Points

VideoStudio doesn't let you create scene markers, or chapter points, in the editing process. Instead, you need to click the Share tab along the top and then click Create Disc on the left to open the disc-creation window. Click Add/Edit Chapter on the left to open a chapter-creation window. Move the playhead under the preview window at the start of your first scene, and then click the Add button on the right. Do this for all of your video's scenes. Click OK when you're done.

Step 2: More Media

Add any other videos that you want included on your DVD into this window, and create scene markers for them as in Step 1. When you're finished, click Next in the lower-right corner.

Step 3: Menu Controls

Select All from the pop-up window in the upper-right corner to see all of VideoStudio's menu templates. Use the controls below the preview window to add a new background image, add background music, or set how long the motion menus play. Click any text in the preview window to edit it.

Step 4: Fine Adjustments

To get the menu looking exactly the way you want, click the Customize button in the bottom-right corner (see Figure 19-7). You can set or change anything about the menu, including how many buttons display, and how the buttons look. When you're satisfied with how the menu and submenus look, click Next.

Step 5: Preview Your Creation

VideoStudio automatically gives you a preview window so that you can test out how your video looks. Use the onscreen remote to click through your DVD menus. Click Next when you're finished.

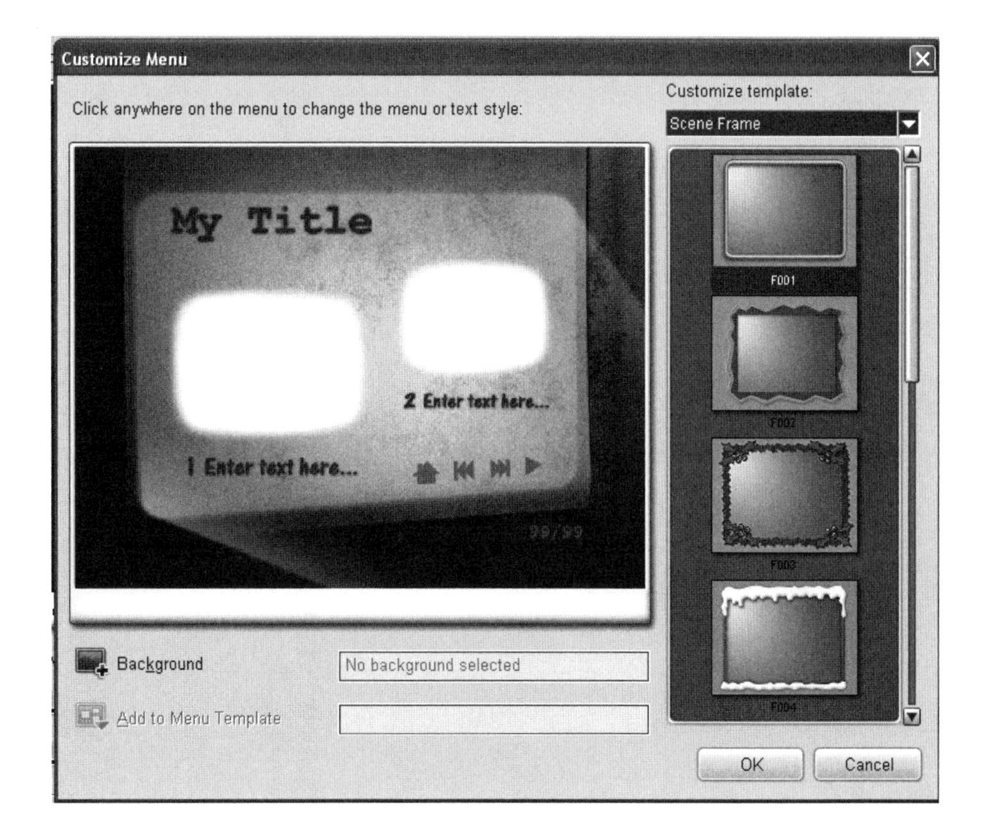

Figure 19-7
VideoStudio's menu customization options

Step 6: Disc Output

The final screen lets you set your disc-burning options. Choose the right disc drive and set your burning speed.

tip *If you're concerned about your computer's ability to burn a disc, click Options at the bottom of the last screen and check the box to perform a test before burning. This will make sure your computer can keep up with the disc-burning process. If your computer fails the test, set the DVD burning speed to something slower and try again.*

With a little time, you can create DVD menus that look just as great as the videos on the disc. Just pick the right theme and finesse the customization options; you'll impress your viewers from the very start.

Part III

Multimedia

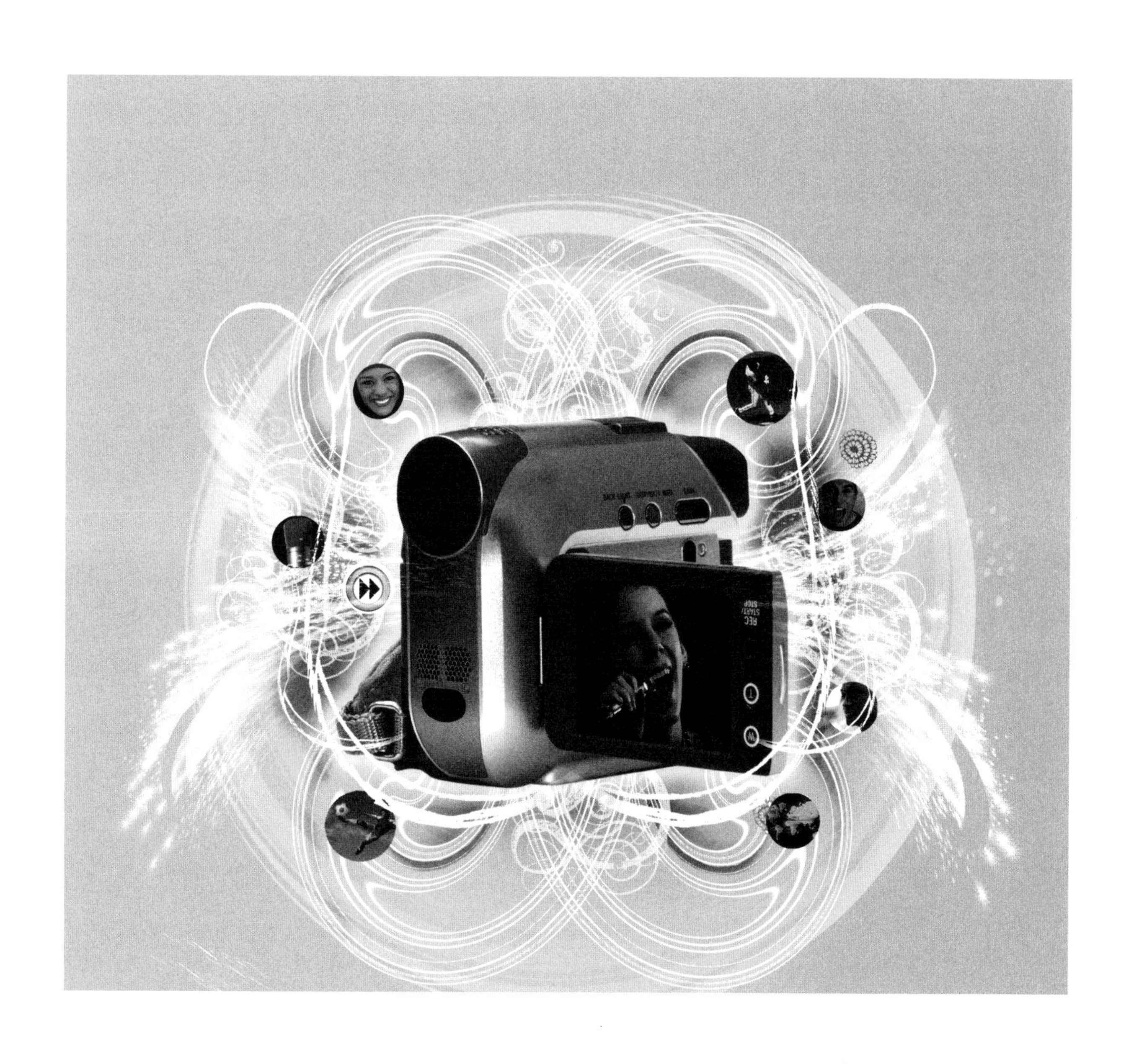

Project 20

Connect Your TV and PC to Make a Digital Movie Theater

What You'll Need

- **Home computer**
- **Television set**
- **Cable, streaming device, or portable music player with TV dock**
- **Cost: $50 to $400**

In the previous projects in this book, we've looked at creative ways you can use your video camera, and technical skills and tricks you can use during editing. Now we're going to turn our attention to projects that go beyond typical video camera shooting. Home video isn't just footage that you shoot and edit yourself. In this age of convergence, not only do you have video on your computer, portable music player (or should that be "portable media player"?), cell phone, smartphone, and, of course, television, but now you can share video between devices. We're rapidly getting to a time when content is freely movable from place to place, so that you can enjoy it wherever and however you wish.

For the time being, though, you'll need to put just a bit of work into sharing your content. That leads us into this project, which explains how to share the video on your computer with your television. Yes, you can burn a DVD, as we've talked about in previous projects, but maybe you want your video to be more fluid than that. Perhaps you prefer to keep your video library on your computer and you don't want to burn everything to a disc. Or perhaps you want the option of viewing your videos immediately on the biggest TV set in the house. If so, you'll appreciate having your computer and television networked together. It's useful for more than just viewing your own creations: you can turn your computer into a digital video recorder (DVR) and save shows (more on that in Project 21), or use your computer to purchase and download shows that you can then stream to your television. Once you have the two linked, you'll find it's a tremendously useful way to live.

Networking with Windows Vista or Windows XP Media Center

Buying a Windows computer with media center features is one of the simpler ways to network your home entertainment center, although few people who have the operating system bother to do so. That's too bad, because you can easily connect a few cables or add a TV tuner to make your entertainment viewing far more flexible.

Microsoft's new Vista operating system comes in several editions, and three of them—Vista Home, Premium, and Ultimate—contain Media Center features. That's a nice switch from Windows XP, which had only one edition that offered Media Center features. While the Vista editions have an improved Media Center interface and new options, the process of making the connection between your computer and TV is the same as in the Windows XP version of Media Center.

Media Center gives you an easy visual way to browse through all the media (movies, photos, and music) on your computer so that you can make a selection. Media Center PCs also come with a remote control, so that you can make your picks from the couch. If you've ever seen a tech magazine refer to a "10-foot interface," it means an options screen that uses big icons and big text, so that you can easily read it from the couch.

Media Center PCs often come with TV tuners, so that you can view and record television programs directly on your computer. To do so, you'll still need to add an antenna or, preferably, a cable connected to your cable programming box or satellite programming box.

More fitting for our project is that Media Center will let you stream content from your computer to your television, where you can enjoy it on a larger screen size and at a better screen quality. Using Media Center is a wired option, so it's a much easier choice if your computer and primary television are in the same room. The idea behind it was to make the computer just another part of your home entertainment system. If your computer is in a different room, you'll want to look into wireless options.

The easiest way to connect your Media Center PC to your TV is with an S-video cable (see Figure 20-1). This is a good-quality connection and it uses a port found on just about every television. Follow these steps to make the connection:

1. If you don't have an S-video cable, visit a local computer shop and pick one up. Don't worry, it won't cost much.
2. Connect the cable to the S-video ports on the back of your computer and your television. You'll need a long cable if they're across the room from each other.
3. Turn on your TV and select PC as the input source. Press the Media Center button on your computer's remote control to call up the Media Center interface.

Figure 20-1
An S-video cable

note *When you make this kind of connection, you're streaming the video to your television, but not the audio. You'll want to plug good-quality speakers into your computer for strong home theater sound.*

There are other types of cables you can choose to connect your PC and TV, depending on the ports available. You can choose an RCA or composite cable (which has red, white, and yellow jacks and delivers audio), a component video cable (which had red, blue, and green jacks), a Video Graphics Array (VGA) cable, a Digital Visual Interface (DVI) cable, or a High-Definition Multimedia Interface (HDMI) cable. HDMI delivers the highest quality, but you won't find it on every computer or TV set just yet. If your TV has the HDMI port, you might want to install an HDMI PCI card on your computer to benefit from the higher quality.

Network Streaming Devices

A more elegant solution to the problem of sharing video is to connect a network streaming device to your television. To do so, you must already have a home network. Once you have a network, you'll find that it's useful for a lot more than just letting two or more computers share a high-bandwidth Internet connection.

With a network streaming device, you install software that comes with the device onto your computer, and then hook the device up to your television (there are also audio-only streaming devices that connect to stereo systems, but we'll focus on the video options). When you turn on the device, you can stream images, videos, and songs from your computer to the television.

Here are a few devices to consider before you make a purchase:

- **HP MediaSmart TV** A flat-panel LCD television that can stream media from your computer.
- **Acoustic Research Digital MediaBridge DMP3000** A networked box that connects to your television and quickly streams media from your home's computers.
- **D-Link MediaLounge DSM-520** A well-priced option for simple media streaming.
- **Microsoft Xbox 360** It's not just a game system; it can also stream content from networked Windows computers.
- **Apple Apple TV** Perhaps the easiest solution of all, the Apple TV (see Figure 20-2) lets you stream any media stored in iTunes. It works with both Macintosh and Windows computers.

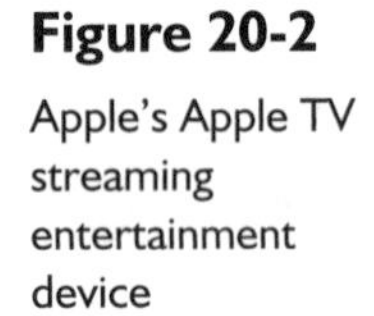

Figure 20-2

Apple's Apple TV streaming entertainment device

Nintendo Wii

Unlike the Xbox 360, mentioned in the previous section, the Nintendo Wii can't stream directly from computers on your home network. But if you have a home network, you can use this simple workaround:

1. Sign up for a free account with Orb (www.orb.com) and download and install the necessary software. Orb is a free service that lets you view the content stored on your computer on any device with Internet access, including cell phones.
2. Turn on the Wii, go to the Shop channel, and download the free Opera web browser. Be sure your Wii is configured to work with your home network.
3. Go to the Wii's Internet channel and use the Opera browser to go to mycast .orb.com to access your account. If your computer is turned on, you'll be able to access any of the media content stored on it.

Orb has created a new interface for Wii users, so that it's easy to browse content from the couch (see Figure 20-3).

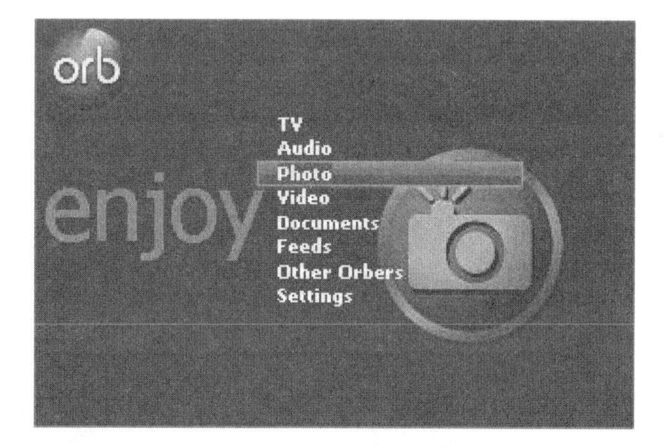

Figure 20-3
Orb's interface for Wii users

Apple iPod

The simplest and lowest-degree-of-difficulty option is to use your iPod as a go-between to carry your videos from your computer to your television. Videos will be compressed to a smaller size, for the iPod screen, so they won't have perfect focus on a television. Still, you'll find that they look surprisingly good. This solution has the advantage of putting your videos on your iPod, so that you can watch them anytime, anywhere.

Here's how to get content onto your iPod:

- **Converting with Apple iMovie** If you're editing in iMovie, choose Sharing | iPod and your video will be converted to the correct format and loaded into iTunes. It will then synch to your device the next time you plug in your iPod.
- **Converting using Adobe Premiere Elements** Choose Export | Adobe Flash Video, Others. You'll get a series of presets, two of which are for the iPod. Select the option for high-quality video, if you have room on the iPod.

Getting your videos onto your iPod is just half the battle. You'll also need to play them on your television. Consider one of these devices, both of which attach to your TV and play content from a docked iPod:

- DLO HomeDock Deluxe (www.dlo.com), shown in Figure 20-4
- Keyspan TuneView for iPod (www.keyspan.com)

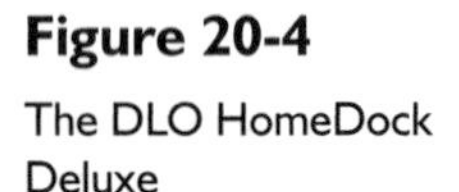

Figure 20-4

The DLO HomeDock Deluxe

Before the age of digital media, people had to set up a white projector screen to watch their home movies. With just a little bit of work, you'll be able to view yours on any television in the house.

Project 21

Save Your Favorite TV Shows

What You'll Need

- **Home computer**
- **TV tuner card**
- **Editing software**
- **Cost: $50 to $350**

Maybe the home video you'd like to record isn't your kid playing softball, your sister's wedding vows, or your baby's first steps. Maybe you'd like to record a whole season of Mets games or the next season of *24*. Turning your computer into a TV recording and editing powerhouse might sound tricky, but there's not much to it. When you're done, you can watch and record all the shows you want, plus you can even remove the commercials before you archive your shows.

In our digital age, there are a few ways to record your shows in digital quality, but the way you're about to learn—installing a TV tuner card—is the best way. Sure, you can record programs with any generic digital video recorder (DVR), but most of them don't let you archive your recordings. You could get a direct-to-disc DVD recorder, which is an easy solution, but that saves the commercials as well, and who wants commercials? You could also get TiVoToGo and burn your saved programs on your computer, but it's an expensive solution, and it doesn't allow you to edit out commercials either. No, for price, ease, and features, installing a TV tuner card is the way to go. If it still sounds too difficult, just follow the easy steps in this project.

TV Tuner Cards

A TV tuner Peripheral Component Interconnect (PCI) card is the ideal way to capture standard TV broadcasts to your computer. With one installed, your PC essentially becomes a TV whenever you want to watch programs. You can even watch shows in a small window while you do your work or surf the Internet with the rest of the screen. Plus, most cards come with software that lets you record what you're watching and schedule recordings, and another application that lets you edit your recordings.

Do a little research online and read some reviews before you buy a card. Hauppauge (www.hauppauge.com) is a good place to start, as it's been making TV tuner cards for many years and has a good reputation. Higher-end programs will let you tune and record radio, and send TV programs out to a TV so that you can watch them on a bigger screen.

Step 1: Pull the Plugs

After you've purchased a TV tuner PCI card, start by unplugging the power cord, monitor cord, and any other cables running into your computer.

Step 2: Turn the Screws

You'll need to open the case, so remove any screws holding the shell or the side panel in place. Put them in a safe place, because they're easy to lose. Slide the panel off. Touch something metal, such as a file cabinet or a large appliance, to dissipate any static electricity that could damage your computer's internal components.

Step 3: Look for an Open PCI Slot

Tip your computer sideways and lay it flat, so that you're looking down into it, as shown in Figure 21-1. You'll need a free PCI slot for your card. PCI slots are long, thin grooves that your PCI card will fit into. You'll need to have at least one slot free to install your card. Remove the narrow metal panel next to the slot, to create an opening on the back of your computer next to the PCI slot you'll use.

Step 4: Get a Good Fit

Line your TV tuner card up over the PCI slot and push it down into place. Try not to wiggle it to get it to fit, as that could damage the card. Instead, push it down gently but firmly. The card's ports will fit in the opening you created when you removed the metal panel.

Step 5: Put It All Back Together

Return the side panel or the computer case and screw it back into place. Turn your computer upright and reattach the cords you removed in Step 1.

Step 6: Connect the Cable

Your TV tuner card will have a coaxial cable port, so that you can plug in the coaxial cable that delivers your cable or satellite programming (see Figure 21-2). You can also

Figure 21-1
Look for a free PCI slot in which to place your TV tuner card.

plug in an antenna, if you don't have a pay service. You probably won't want to unplug the TV service from your actual TV, so you'll need to split the cable into two. Visit your local computer shop and pick up a cable splitter and a length of coaxial cable.

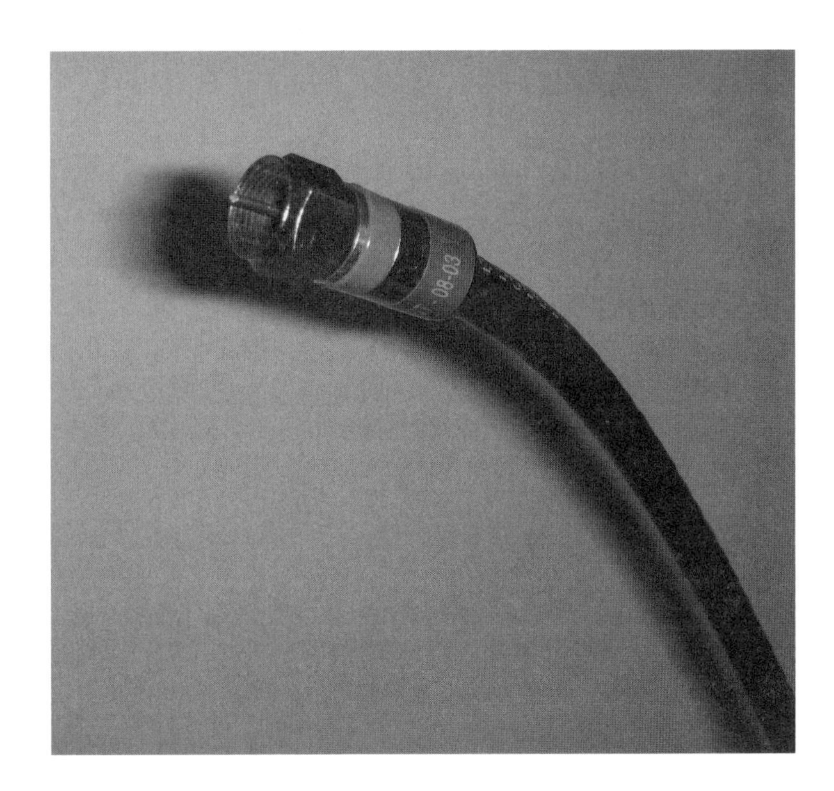

Figure 21-2
Plug your television service's coaxial cable into the back of your computer.

Step 7: Install the Software

Power up your computer and install the software that came with your TV tuner card. Once you have it running, you should find that viewing and recording television programs is easy.

Look for a TV tuner card that comes with a remote, so you can enjoy programming from the comfort of your couch.

External Cards

If you don't have an open PCI slot or if you have a notebook (or if you really don't like the idea of opening your computer), there are external TV tuners on the market (see Figure 21-3). These typically connect with a USB cable and deliver all of the same features as an internal TV tuner card. The downside is that image quality is often not quite as good, although pricing is about the same.

Figure 21-3
External TV tuner cards offer an easier way to access TV programs on your computer.

Editing

You'll probably find that the software bundle that comes with the TV tuner card lets you record TV shows, create a schedule of future and regular recordings, and edit your videos. The editing software is likely to be no-frills, so you might prefer using your standard video editing program. The TV tuner card will record shows in common formats, so you should be able to edit out commercials using any software.

Step 1: Load the Video

Load a recorded TV program into your video editing program.

Step 2: Time for Trims

Trim out any extraneous material at the beginning or end of your clip by moving the playhead to the very start or end of your show and then clicking the correct button to split the video. Click the unwanted material and then press DELETE. Next, trim out commercials by positioning your playhead just before and after commercials and then deleting the blocks of commercials.

Step 3: Save It

When you're done with your trims and edits, save your work so that you now have ad-free versions of your favorite shows.

When you're finished, you'll be able to record any show and burn a DVD archive so that you can enjoy your favorite programs whenever you like. Your computer could turn out to be the best TV you've ever owned.

Project 22
Share Your Videos

What You'll Need

- **Digital video camera (any)**
- **Home computer**
- **Movie editing software**
- **Cost: $100**

For most of this book, we've focused on burning your videos to a DVD to either save them for your own viewing or to give them as a gift. But digital media is more fluid than that, and the more video you shoot, the greater the chance that you'll want to share it in other ways. Sharing video files is not quite as simple as sharing digital photographs, because video files are much larger than photo files. You can share your videos in the same way that you share photos—via e-mail—but you need to know a little bit about export options and file sizes. You don't want to attempt to send a full-size copy of your video via e-mail, because it would probably be too large to go through the mail server; and even it if did go through, the recipient might be angry at you for sending such an enormous attachment.

E-mailing your videos isn't the only way to share them. Chances are you've done a little browsing on YouTube, the popular video sharing site. Once you have a completed video, you can create a YouTube account and upload your creation, sharing it with the world. If you're doing creative work, it's a great way to get some exposure and get some feedback from other people. This project will guide you through the steps.

Sharing by E-mail

The secret to successfully e-mailing your videos is to decrease the file size, so that the attachment isn't too large for your recipient's mail client. Most mail servers can handle attachments of up to 10MB, but that doesn't mean you should aim for that size. Smaller is better. The flipside to this is that the movie will display smaller as well. If you export a video for e-mail, it's going to have a smaller resolution (a measurement in pixels of the movie's length and width), and it will have fewer frames per second. The recipient will still get to see the movie, but it won't be as large or at the same level of quality.

We'll use Apple iMovie for our examples in this project, although all video editing programs have similar export options.

Step 1: Select the Settings

Open the Share drop-down menu in iMovie and you'll see that sharing is typically a one-step process. You select the destination for your video, and iMovie does the work of exporting a version to match a list of preset conditions. For example, the e-mail setting calls for your video to have a resolution of 160 by 120 pixels, ten frames per second, and monaural sound (see Figure 22-1). Select this option and iMovie will go to work, giving you a progress bar to show how far it's gotten. When it's done, it will launch your e-mail application and create a new message with the saved video as an attachment. Double-click the attachment to see what it looks like.

Figure 22-1
Apple iMovie's e-mail sharing settings

Step 2: Other Options

If you don't want to send such a small movie, choose the option to save your file as a QuickTime movie, and then choose Expert Settings from the pop-up window. When you click Share, iMovie will first let you set how your movie should be saved. If you're sending your movie to a Windows user and you know they don't have QuickTime loaded on their computer, choose to save your movie as a Windows Video file. You can then choose the quality of your movie by specifying how you'll be sending it. If you're sending your movie to someone who still accesses the Internet over a dial-up connection, choose the dial-up setting. This will ensure that your video isn't too large and won't take too long to download. If size isn't a problem, choose the option for a high-volume broadband connection.

Formatting for an iPod

The Apple iPod isn't the only portable media player on the market, but it's overwhelmingly the favorite, so it's what we'll focus on here. If you've got an iPod that is capable of playing video, you'll enjoy having your movies stored on it so that they're always close at hand. There are two easy ways to convert video for iPod use.

Export to iPod

When you're finished with your movie, open the Share drop-down menu in iMovie and choose the iPod option (see Figure 22-2). This not only puts the video in the right format, but stores it in iTunes so that it's automatically loaded onto your iPod the next time you synch. Click Share to start the conversion.

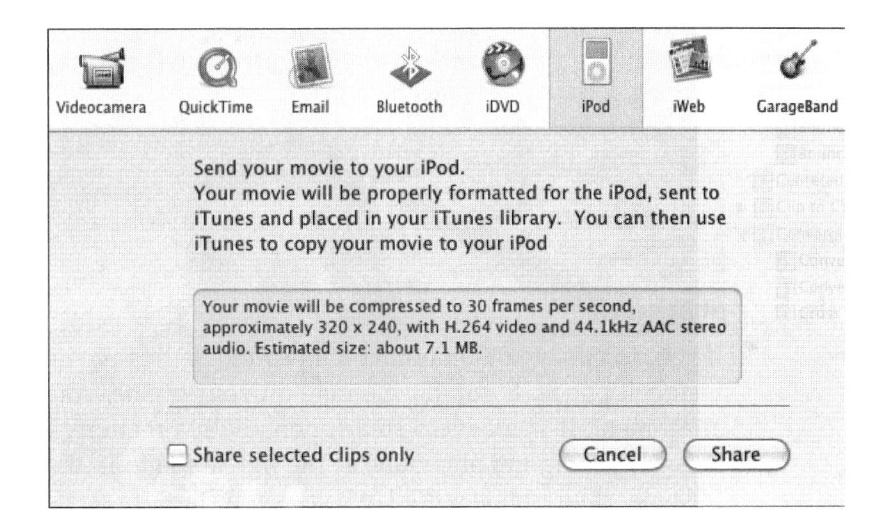

Figure 22-2 Use iMovie's iPod Settings to export a movie to an iPod.

The iTunes Conversion

A second option is to save your movie to your desktop at whatever high-quality setting you want, and then import it into iTunes as detailed in the previous section, by choosing File | Import. To make the file iPod-ready, find it in the iTunes library and either right-click it (if you have a multibutton mouse) or CONTROL-click it. From the resulting pop-up menu, choose the option Convert Selection for iPod. This will convert the file for you, so that it's ready to synch the next time you connect your iPod.

Formatting for a Cell Phone

You can enjoy your movies on a cell phone, but you'll need to view them on a much smaller screen. Choose one of these methods:

Bluetooth Transfer

If you have Bluetooth on your cell phone and your computer, you can send the file without any wires. Make sure your phone is in Bluetooth discoverable mode before you begin. Choose Share | Bluetooth (see Figure 22-3). This will create a much smaller version of your movie, which will look good on a cell phone and won't use too much space.

Figure 22-3
If your computer and phone both have Bluetooth, you can use iMovie to wirelessly transfer your movies.

File Export

If you don't have Bluetooth, you'll need to be creative in how you get your movie to your cell phone. If you have e-mail on your phone, you can send your movie as an attachment. If you have a smartphone with a memory card, you can load the card into your computer and transfer the movie file to it. Be sure you know what video file types your phone will play before you start. If you have a Windows smartphone, convert your movie to the Windows Video format. Make the quality the absolute least you can, which will still look okay on a tiny screen.

Uploading to YouTube

When YouTube (www.youtube.com), the video sharing site, exploded in popularity, it was inevitable that a variety of similar sites would spring up. Some, such as Revver (www.revver.com) and Metacafe (www.metacafe.com), will pay you if your video becomes popular. Truthfully, though, they don't pay much. You're better off simply posting to YouTube, where your videos will have a better chance of being seen by a lot of people.

Step 1: Register for an Account

Before you can upload videos to YouTube, you'll need an account. Click the Sign Up link at the top of YouTube's home page and follow the directions to create a standard account. Doing so is free.

note *Want to post a quick rant online without fussing over editing? YouTube can grab video from a connected webcam so that you can create instant videos.*

Step 2: Upload Your Files

Sign in with your new account and click the Upload Videos link at the top of the screen. You'll be asked to input the title, a short description, and keywords (called *tags*), that describe your video (see Figure 22-4). These tags help people find your movie when searching.

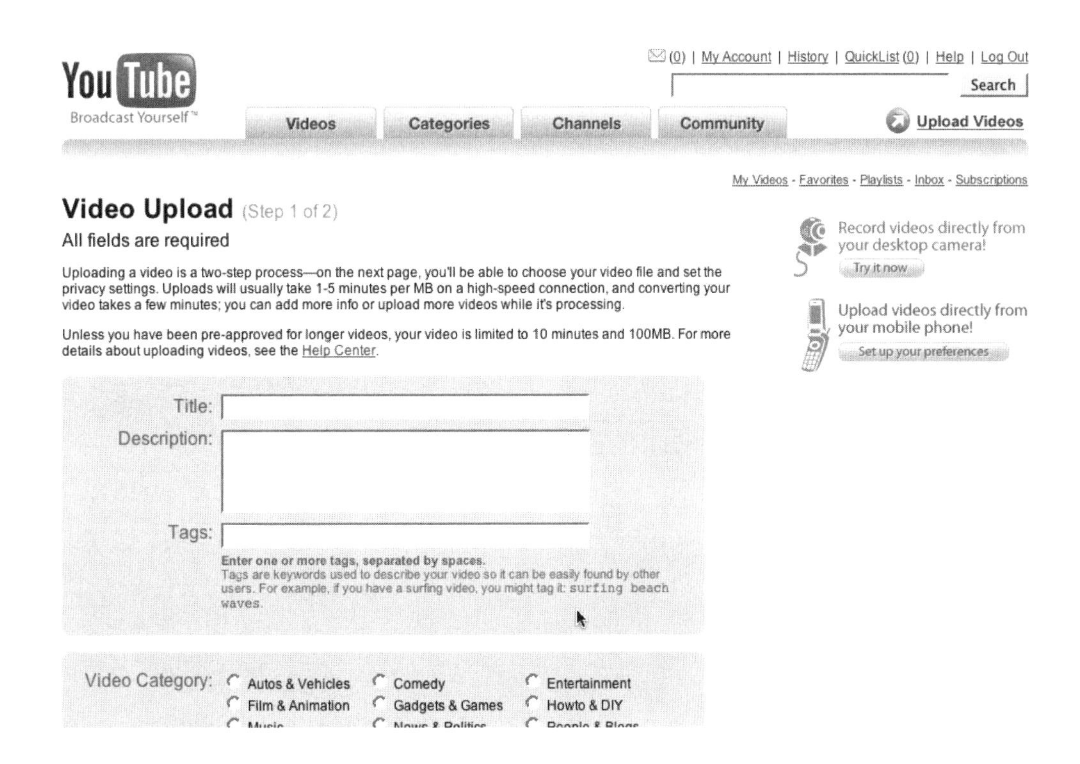

Figure 22-4
YouTube's video upload controls

note *Videos uploaded to YouTube can only be ten minutes in length and no greater than 100MB. Don't worry about format; YouTube can accept all popular video formats. The site automatically coverts videos into the Flash Video format, to save space. As a result, your work won't have the same sharpness online that it has on your computer.*

You'll also select the category and the language of your work. Next, you'll select the file on your hard drive and set whether you want your video to be public or private. If you make it private, you'll need to create a contact list of people who can view it.

tip *YouTube can accept videos up to ten minutes in length with a standard account, but you should keep your clips under three minutes so that your audience won't get bored and tune away. Short clips work better online.*

Gone are the days when showing home video was a major production. Now, with sharing sites and broadband Internet connections, sharing your creations with family and friends is surprisingly easy—and fun, too.

Project 23

Create a Video Podcast

What You'll Need

- **Digital video camera or desktop webcam (any)**
- **Home computer**
- **Movie editing software**
- **Cost: $100**

Would you like to be the star of your own TV show? Ham it up and deliver one-liners on the news of the day? Give lessons on a topic you're passionate about? Or perhaps post a video diary of what's new in your life? Then you should create a video podcast. Also known as vodcasts or vlogs, video podcasts are video segments that you shoot, edit, and post online, and which other people can subscribe to. Your subscribers can then transfer the video podcast to their iPod or other portable media player, or watch it on their computer.

Setting up a video podcast is just a bit tricky—no one has yet come up with a completely simple solution—but this project will make it as smooth as possible. You should know ahead of time that there are many ways to accomplish the same basic task, so if you keep going with podcasting, you may find paid software or services that you like better than the ones included here. This project discusses free tools that make the process fairly simple, but there are other options that deliver more control. For now, start with these instructions; they'll get you started so that you can see if video podcasting is right for you.

Step 1: Getting Started

To be a video podcaster, you need to have four things: a video in a correct format, a place to host your blog, a place to host your video files, and an RSS feed.

How many people are about to flip to an easier project?

Look, it's really not that hard. This project will walk you through it one step at a time.

Step 2: Make the Video

The subject of your video podcast can be anything you like. People have been successful with all types of formats, so your imagination is your only constraint. One of the most popular video podcasts today is Ask a Ninja (www.askaninja.com), where a killing-happy ninja clad in all black answers viewers' questions (see Figure 23-1). The popularity of this site shows that there's an audience hungry for things they've never seen before—as long as those things are done well.

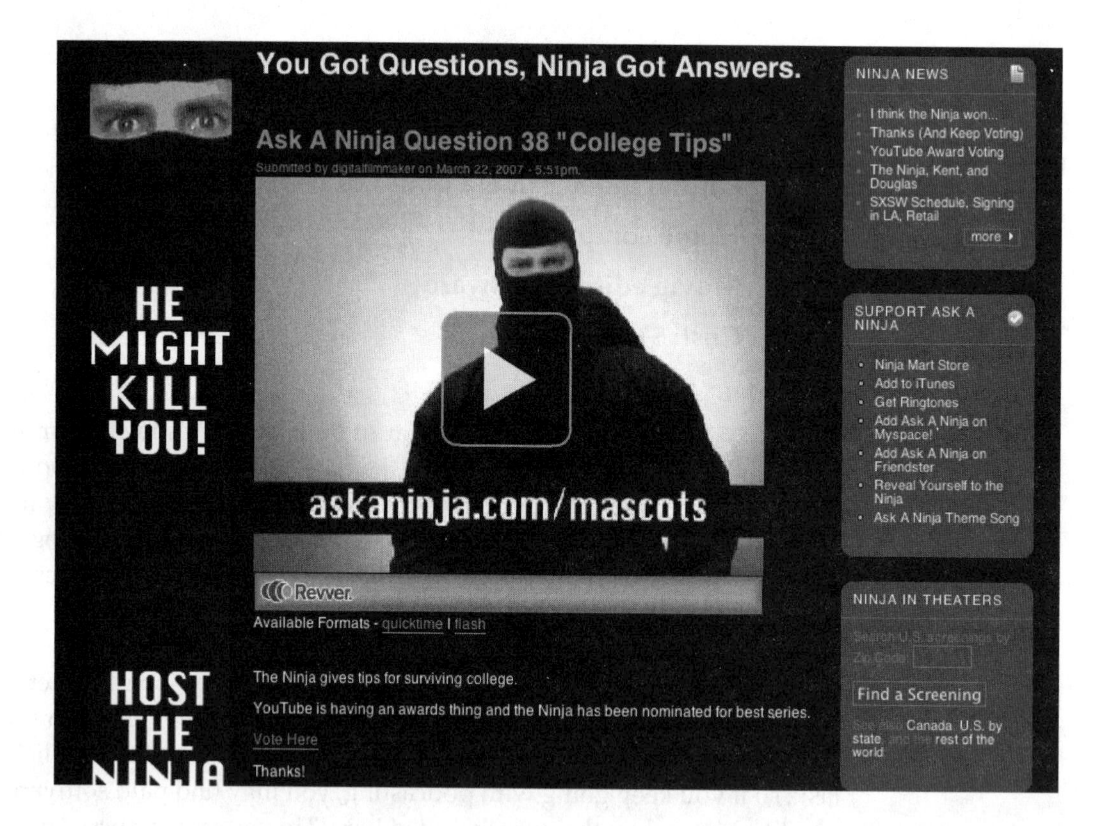

Figure 23-1
Home page for Ask a Ninja

While you can make your podcast pretty long if you want, keep it short at the beginning. Get the feel for podcasting first, before you start pushing yourself. Short video podcasts are easier to watch as well. You don't want to tax your audience's patience. Leave 'em wanting more.

You don't need a handheld video camera to shoot a video podcast. Lots of people create them with desktop video cams. It depends on what your video podcast will look like. If you intend to sit at your desk and talk, a desktop video cam will be fine. Besides, some of today's better webcams—from Apple (see Figure 23-2), Microsoft, and Logitech, among others—can take great-quality video.

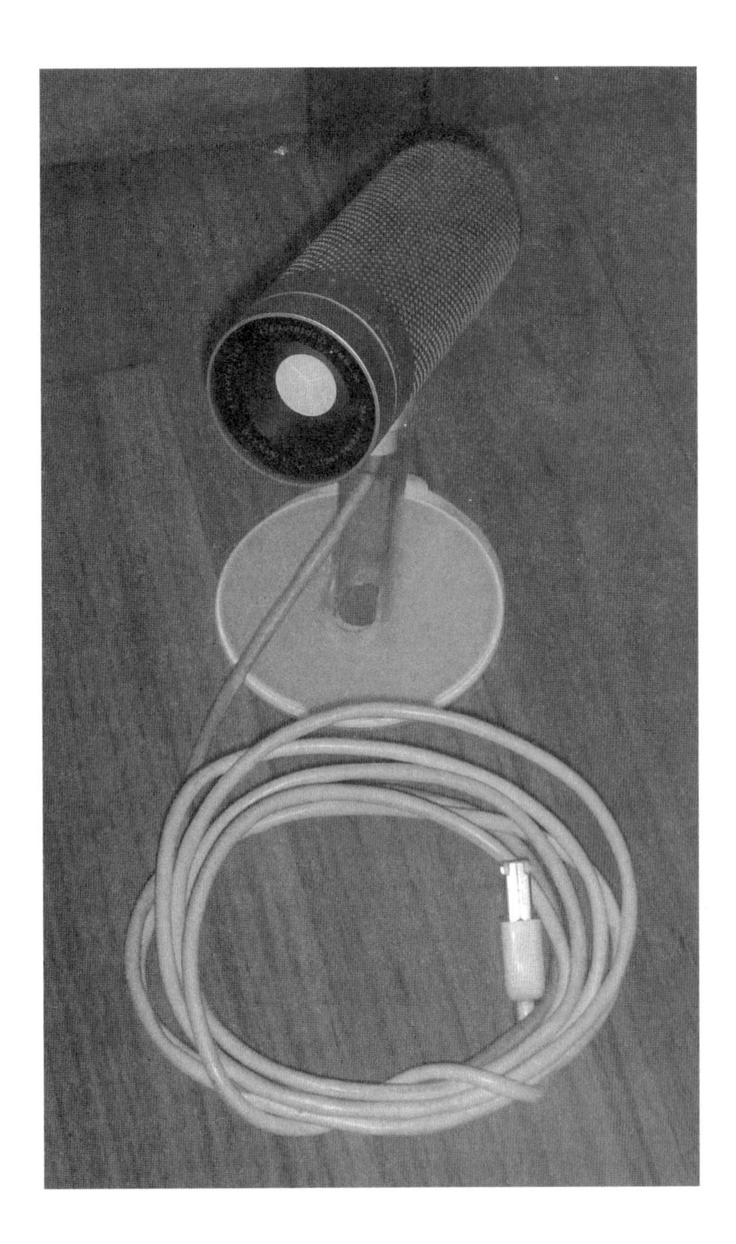

Figure 23-2
The Apple iSight desktop webcam

Once you've shot and edited your work, you need to get it in the right format for podcasting. It should be a 320×240-pixel MP4 or H.264 file. This makes it compatible with the video-enabled iPod, which is how most people will view your work.

- **To export with a Mac** Edit your video with iMovie and choose Share | iPod when you're done. This converts the video for iPod (and puts it in your iTunes directory) so that it's podcast-ready by default.
- **To export with a Windows computer** Click the Export button at the top center of the Adobe Premiere Elements interface and choose Adobe Flash Video, Others. In the new window, select either of the iPod export options.

When you're done, load the video into iTunes (if necessary), select it in the iTunes directory, and then call up the file's information options. Give it a name, set the author, and add cover art.

Step 3: Create a Blog

Before you can have a video blog, you need a regular blog. Don't worry, you don't have to write much in it. For each video podcast, you'll first create a text entry in the blog and then attach the video to the text entry. This is part of how video podcast distribution works.

Go to Blogger (www.blogger.com; see Figure 23-3), a free online blogging service, and create a blog by clicking the Create Your Blog Now link. You'll be guided through the process of naming a blog and choosing a template. Give it the same name as your video podcast.

Figure 23-3
Go to Blogger to create a free blog.

note *You can use any blogging service for this step. Consider Blogger, because it's free and simple.*

Step 4: Find Media Hosting

Blogger won't host your video files, so you need to find someplace that will. If you have an account with an Internet service provider (ISP) for Internet service, you might be entitled to free online storage. Ask your ISP whether your account includes free online storage and, if so, ask how to upload files. If you have a .Mac account, online storage is included. If you own a domain name and run a web site, you can use that online storage.

If none of that applies to you, sign up for a free account with Blip.TV (www.blip.tv; see Figure 23-4), a site for creative people that offers free storage and free bandwidth. That's a pretty nice offer.

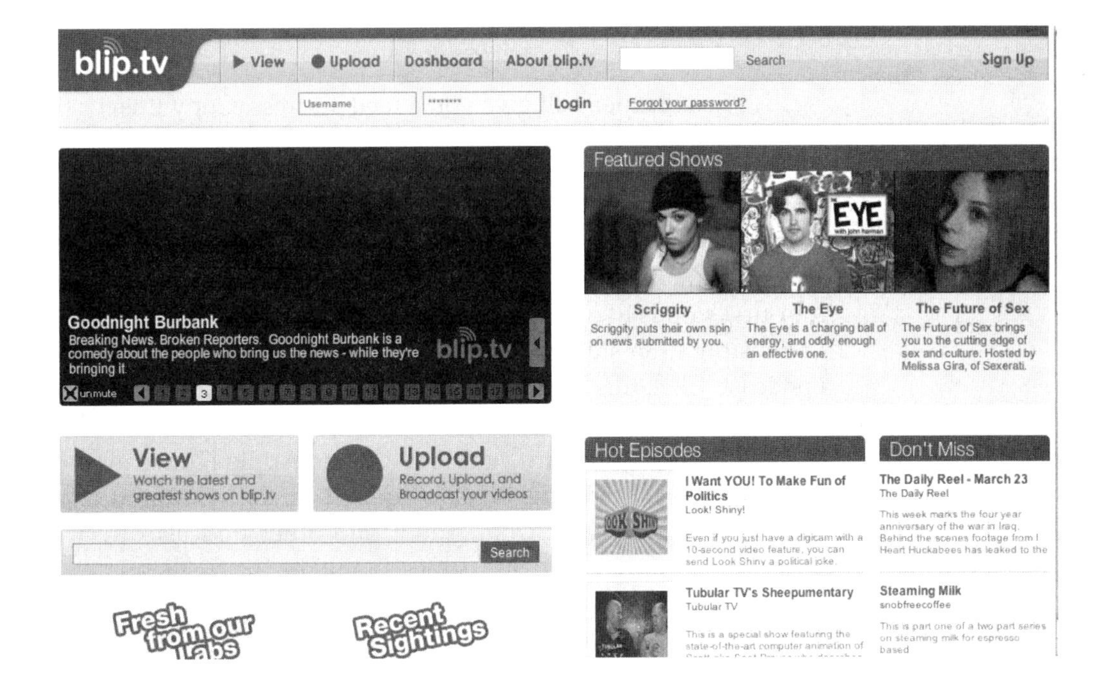

Figure 23-4 Blip.TV will give you free hosting for your videos.

When you've signed up for an account, upload your first video to the site. Right-click the video to get its URL. Keep that handy.

Step 5: Make an Entry

You're all set now to create your first blog entry. Go back to your Blogger account, open your Blogger Dashboard, and click to create a new post. You can make it a long entry, or just tell people to look at your video. In either case, select some text and make it a link to the video by highlighting it and clicking the linking tool in the Blogger control panel. For the link, paste in the URL you got for your online video in the previous step.

Step 6: Get an RSS Feed

Your video is now online, so you're almost done. The next step is to get an RSS feed that points to your video. "RSS" stands for Really Simple Syndication, and it's the way for people to subscribe to your video podcast. The RSS link will let people (actually, the podcast software they're using) know when there's a new entry to your video podcast. Then, it will download the new entry automatically. To get an RSS feed:

1. Copy the URL for the main page of your blog.
2. Go to a free web service called FeedBurner (www.feedburner.com). Paste your URL into the big text field in the middle of the page and click the box that says "I am a podcaster," then click Next. The next screen will ask you to title your feed, customize the feed address, and set up a password. Click Activate Feed when you're done.
3. FeedBurner will give you a feed address. Copy this for the next step.

Step 7: Spread the Word

Now that you're set up with a video podcast and a feed, you need to spread the word. Your first stop is to open up iTunes and open the iTunes Store. Click the podcast link on the store's left side. Scroll down the podcast page and you'll see a link to submit a podcast. Click this and you'll be prompted to add information for your video podcast, starting with the feed address that you received from FeedBurner.

Podcasting's popularity exploded when Apple iTunes made it easy to subscribe to them. iTunes is still the biggest podcast directory around and it's where most of your viewers will come from. You'll find other podcast directories where you can submit your information, but iTunes should be the first stop.

See, that wasn't so hard. With a little effort, your video podcast will be ready to go and you'll be attracting a worldwide audience.

Project 24

Create an Archive That Will Last for Years

What You'll Need

- **Home computer**
- **Movie editing software**
- **Archival-quality DVDs**
- **Cost: $130**

Our last project is short and sweet, but it could be the most important one in this book. If you've read this much, you obviously care about shooting great video, so spend a little time and money protecting your work. If you burn your videos to DVD, the most practical solution for long-term storage, you'll be horrified to learn how short a time some discs last. When you burn a movie to a DVD, you probably assume that the disc will last at least as long as you will. But if you're using bargain-priced DVDs, they could begin to disintegrate in as little as ten years. (Many cheap ones won't burn at all.) No DVD is perfect, and those precious home movies will eventually need to be transferred to a new storage medium, but you can make them last much longer if you use the right media. In this project, you'll learn which discs to buy and how to care for them correctly. You'll be surprised at the conditions that can affect a DVD's lifespan. Following this advice will cost you a little more at the checkout counter, but you'll be buying peace of mind.

Choosing Archival Media

There's only one step to this project: buy archival-quality DVDs (see Figure 24-1). That's it. These DVDs are made to a much higher standard, and will resist breaking down much better than average discs. The term "archival" isn't just marketing hype. A well-cared-for archival DVD can last 200 years (good-quality nonarchival discs last 50 to 100 years). By then, your descendents will have had more than enough time to transfer your videos to some new and as of yet undreamt of medium.

Figure 24-1
Buy archival-quality DVD+R discs for the best long-term storage.

The catch is that archival-quality discs cost more...not a lot more, but perhaps double the price of standard blank DVDs. Pay the difference. If it's something you want to save, it's worth the money.

Tips for DVD Care

Buying the best blank DVDs is essential, but there's more you can do to ensure the quality of your discs. Follow these tips to keep your DVDs in the best condition:

- Always handle your discs by the edges. You've probably learned this with CDs, so perhaps you're already doing it. You don't want to touch the data side of a DVD.
- Don't use an adhesive label on a DVD, such as those shown in Figure 24-2. Lots of people create one after they burn a disc, but applying an adhesive label can warp the disc.
- Store your discs vertically and keep them in a jewel case. The jewel case seems like a wasteful design, since it takes so much room, but it actually serves a purpose. It holds the DVD or CD in space so that it isn't touching the front or back of the case. That protects the data on the disc. Store your discs vertically, because storing them horizontally causes them to bend over time.

Figure 24-2

Adhesive disc labels can warp a DVD or CD over time.

- Don't expose your DVDs to direct sunlight. Doing so can actually break down the media stored on it. They should always be kept in a dark place.
- Don't use RW ("rewritable") discs for long-term storage. They're not good for archiving. Instead, buy DVD+R ("recordable") discs, which last the longest.
- Buy brand name discs. You get what you pay for, and no-name bargain DVDs are no bargain. For archiving, Taiyo Yuden, Maxell, Sony, TDK, Pioneer, and Verbatim are all top names. Many users consider Taiyo Yuden to be the best.
- Don't write on top of your disc with a pencil or ballpoint pen. Use a felt tip. Obviously, never write on the data side at all.

If you've spent time creating a video, spend a little money insuring that it will last. You'll be able to watch your movies throughout your life, and your kids will be able to watch them throughout their lives, too.

Index

A

B

C

I

K

L

M

N

O

P

T

U

V

W

Y